UNIVERSITY OF STRATHCLYDE

ISNM
International Series of Numerical Mathematics
Vol. 127

Managing Editors:
K.-H. Hoffmann, München
D. Mittelmann, Tempe

Associate Editors:
R. E. Bank, La Jolla
H. Kawarada, Chiba
R. J. LeVeque, Seattle
C. Verdi, Milano

Honorary Editor:
J. Todd, Pasadena

Optimal Control of Soil Venting: Mathematical Modeling and Applications

Horst H. Gerke
Ulrich Hornung[†]
Youcef Kelanemer
Marián Slodička
Stephan Schumacher

Birkhäuser Verlag
Basel · Boston · Berlin

Authors:

H.H. Gerke
Center for Agricultural Landscape and
Land Use Research (ZALF)
Department of Soil Landscape Research
Eberswalder Str. 84
D-15374 Müncheberg, Germany

Y. Kelanemer
5252 St Hubert # 445
Montreal H2J3Z6
Quebec, Canada

M. Slodička
Institute of Applied Mathematics
Faculty of Mathematics and Physics
Comenius University
Mlynská dolina
84215 Bratislava, Slovakia

S. Schumacher
Bayerische Landesbank
Brienner Str. 24
D-80333 München, Germany

1991 Mathematics Subject Classification 73K40, 76T05, 86A32

A CIP catalogue record for this book is available from the Library of Congress, Washington D.C., USA

Deutsche Bibliothek Cataloging-in-Publication Data

Optimal control of soil venting: mathematical modeling and applications / Horst H. Gerke ... – Basel ; Boston ; Berlin : Birkhäuser, 1999
　　(International series of numerical mathematics ; Vol. 127)
　　ISBN 3-7643-6041-0 (Basel ...)
　　ISBN 0-8176-6041-0 (Boston)

This work is subject to copyright. All rights are reserved, whether the whole or part of the material is concerned, specifically the rights of translation, reprinting, re-use of illustrations, broadcasting, reproduction on microfilms or in other ways, and storage in data banks. For any kind of use whatsoever, permission from the copyright owner must be obtained.

© 1999 Birkhäuser Verlag, P.O. Box 133, CH-4010 Basel, Switzerland
Printed on acid-free paper produced of chlorine-free pulp. TCF ∞
Cover design: Heinz Hiltbrunner, Basel
Printed in Germany
ISBN 3-7643-6041-0
ISBN 0-8176-6041-0

9 8 7 6 5 4 3 2 1

Contents

Preface	ix
Abbreviations	xi
Notations	xi
List of Figures	xii
List of Tables	xiv

1 Introduction
  1.1 Background and Problem 1
  1.2 Objectives 4
    1.2.1 Mathematical and Numerical Objectives 4
    1.2.2 Objectives of the Optimization 5

2 Modeling Soil Venting
  2.1 Physical Basis of the Venting Technique 7
  2.2 Simulation Models 9
    2.2.1 Soil Gas Transport Models 9
    2.2.2 Phase Mass Transfer Models 10
    2.2.3 Multi-phase Permeability Models 11
    2.2.4 Biological Decay 11
    2.2.5 Measurement Configuration 12
    2.2.6 Heterogeneous Media 12
    2.2.7 Structured Soil 12
    2.2.8 Mathematical Aspects 13
    2.2.9 Remarks on Preferential Gas Flow in Structured Soils 14

3 Stationary Problem and Optimal Control
  3.1 Basic and State Equations, Objective Function 16
    3.1.1 Air Flow 17
    3.1.2 Contaminant Transport 18
    3.1.3 Optimal Control 19

	3.2	Mathematical Analysis, Simplifications	20
		3.2.1 Dimensionless Form	20
		3.2.2 Approximation of Sinks by Dirac Functions	24
		3.2.3 Justification of a Leakage Term	25
		3.2.4 Streamline Method	30

4 Well-Posedness and Optimality

4.1	Air Flow in a Domain with Holes	33
4.2	Air Flow with Dirac-Type Sinks	34
4.3	Non-smooth Transmissivity at Active Wells	37
4.4	Contaminant Transport in a Domain with Holes	40
4.5	Differentiation of the Transport Equations	42

5 Optimization of Simple Well Configurations

5.1	One Single Active Well	45
	5.1.1 Constant Volatilization on a Disk	45
	5.1.2 Bell-Shaped Volatilization	46
5.2	Two Active Wells	52
5.3	Configurations with a Few Wells	53

6 Estimating the Coefficients

6.1	Pedotransfer Model	59
	6.1.1 Grouping of Soil Substrates	59
	6.1.2 Grouping of the Texture of the Fine Soil Material (equivalent grain diameter smaller than 2 mm)	60
	6.1.3 Estimation of the Cumulative Particle-Size Distribution	60
	6.1.4 Graphical Interpretation of the Particle-Size Distribution	60
	6.1.5 Estimation of the Hydraulic Conductivity at Pore Water Saturation	62
	6.1.6 Permeability of the Refilled and Rubble Material	64
	6.1.7 Analytical Interpretation of the Particle-Size Distribution	65
	6.1.8 Residual Water Saturation	66
	6.1.9 Hydraulic Parameter Functions	67
	6.1.10 Relative Permeability, Saturation-Pressure, and Relative Permeability Relations	67
	6.1.11 Pedotransfer Functions, Pressure-Saturation Relations	68
	6.1.12 Discussion	70
6.2	Spatial Variability and Uncertainties	72
6.3	Ordinary Kriging	73
6.4	Generation of a Random Field	77

	6.5	Parameter Identification of Kertess	85
		6.5.1 Calibration of the Transmissivity T, the Leakage L and the Boundary Leakage l	88
		6.5.2 Second Calibration of the Transmissivity T	91
		6.5.3 Results of the Calibration of the Air Flow Parameters	92
		6.5.4 Calibration of the Volatilization Coefficient V	92
		6.5.5 Numerical Minimization Method	96
		6.5.6 Results of the Calibration of the Volatilization Coefficient	96
7	**Numerical Methods and Optimization**		
	7.1	Gas Flow Field	100
		7.1.1 Linearity of the Flow Field w.r.t. the Control Variable	100
		7.1.2 Mixed Variational Formulation	101
		7.1.3 Discrete Variational Formulation	102
	7.2	Contaminant Transport	103
	7.3	Integrating the Extraction Rate	105
	7.4	Optimization Algorithm	106
	7.5	Programming Aspects	109
8	**Applications**		
	8.1	Kertess	111
		8.1.1 Optimizing the Extraction Rates	115
	8.2	Kirchweyhe	121
		8.2.1 Optimizing the Extraction Rates	123
9	**Stochastic Optimization**		
	9.1	Monte Carlo Optimization	127
	9.2	Sensitivity Analysis	134
10	**Discussions and Conclusions**		
	10.1	Time Dependent Problem	138
	10.2	Optimization of Well Positions in Large Scale Applications	138
	10.3	Further Aspects of Stochastic Optimization	139
		10.3.1 Solution Strategies	139
		10.3.2 Solution Strategies	140
Bibliography			141
Index			151

Preface

No more than about 10 years ago, it seemed to be unrealistic to simulate the complex flow, transport, and transformation processes of organic compounds in multi-phase systems of water-unsaturated soils (Schwille [Sch84]). Numerical solutions of comprehensive models consisting, for instance, of a system of three partial differential equations and chemical reactions based on local thermodynamic equilibrium assumption (Abriola and Pinder [AP85a]), were at that time regarded as far too computational expensive; optimizations seemed to be by far not practicable using such numerical models. However, even more complex mathematical models have been developed since then, like, for example, by considering the temperature-dependency of processes and non-equilibrium phase-transfers, or the increase of the number of components and spatial dimensions involved (e.g., Bear and Nitao [BN92]). Those models require efficient numerical solution algorithms adopted to the possibilities of the latest computer developments.

The book describes the work of a three years research project funded by the The Federal Ministry for Education, Science, Research and Technology (BMBF), Grant Number 03-HO7BWM.

The idea for the project was developed by Prof. Dr. Ulrich Hornung, Dr. Dirk Stegemann, and Dr. Horst H. Gerke in late 1992 after intensive discussions about the BMBF-opened program (*Die anwendungsorientierte Verbundprojekte auf dem Gebiet der Mathematik des Bundesministeriums für Bildung, Wissenschaft, Forschung und Technologie*) on applied mathematics in industrial areas. Dr. Stegemann worked at that time for the company GEO-data, Garbsen near Hannover. GEO-data was engaged in the remediation of the heavily contaminated former Kertess-industrial area which was a pilot-BMBF-project on VOC-contaminated sites. Dr. Stegemann envisioned that such a project could improve remediation efforts developed at that site, further development and use for future contaminated sites.

The BMBF-program offered the opportunity for interdisciplinary work in which applied mathematical developments are to be introduced into technical engineering fields in a way that the problems defined by companies can be solved more efficiently. The basic idea was to involve industrial partners, such that the results obtained in the projects would be immediately used and further developed by the companies.

The manuscript was initiated by Prof. Dr. Ulrich Hornung who laid foundation of the basic structure of this book. The whole text is divided into 10 chapters. The Introduction explains the motivation, ecological reasons and the necessity of the modeling of soil venting. The physical basis and different types of mathematical models are described in Chapter 2. The next chapter deals with a stationary (air flow and contaminant transport) problem as well as with an optimal control problem. We discuss here some simplifications concerning the well modeling. Chapter 4 has mathematical character. It is devoted to the study of the existence and uniqueness for the air flow and contaminant transport problem. Some simple well configurations are discussed in Chapter 5. Estimating of coefficients plays an important role in the modeling. Chapter 6 shows the way how to obtain, interpolate and calibrate the functions describing the transmissivity and contamination. The numerical and computational part is discussed in Chapter 7. An important part of the mathematical modeling is its verification in praxis. The models have been applied to the two real remediation sites (Kertess and Kirchweyhe in Germany). The results of this are shown in Chapter 8. The stochastic aspects and the influence on the results of the optimization are discussed in Chapter 9. Some other ideas and aspects, which have not been studied in this book, are mentioned in the last chapter.

Acknowledgments

This work was supported by the German Federal Ministry for Education, Science, Research, and Technology (BMBF) under Grant No. 03-HO7BWM.

We acknowledge the *Deutsche Bahn AG, Geschäftsbereich Netz, Projektzentrum Nord, Joachimstraße 4/5, 30159 Hannover, Germany*, the owner of both remediation sites Kertess and Kirchweyhe for their support and willingness, to submit data and results of other activities which have significantly helped us to succeed in our scientific work.

The authors wish to thank the partner company *GEO-data in Garbsen, Germany*, which supported technical assistance and monitoring data on soil descriptions and chemical analyzes from the Kertess and Kirchweyhe projects, and for many helpful discussions during the duration of the whole project.

We are indebted to *Dr. Dirk Stegemann (Chemical Laboratory Dr. Spefeld, located in Georgsmarienhütte, Germany)* for his ideas and assistance in this work.

Abbreviations

- BVP = **B**oundary **V**alue **P**roblem
- NAPL = **N**on **A**queous **P**hase **L**iquid
- ODE = **O**rdinary **D**ifferential **E**quation
- PDE = **P**artial **D**ifferential **E**quation
- SVE = **S**oil **V**apor **E**xtraction
- VOC = **V**olatile **O**rganic **C**ompound

Notations

- C $\left[\frac{kg}{m^2}\right]$ is the 2-D concentration of a given VOC in the soil
- C_S $\left[\frac{kg}{m^2}\right]$ is the 2-D concentration of the VOC at the equilibrium. For 1,1,1-Trichloroethen we have from [BHM89a], $C_S = 0.690 \frac{kg}{m^3}$
- D $\left[\frac{m^3}{s}\right]$ $(8 \cdot 10^{-6} \frac{m^3}{s})$ is the diffusivity of the VOC
- $\mathbf{F}$ $\left[\frac{kg}{ms}\right]$ is the 2-D mass flux density of the VOC
- $\mathbf{f}$ $\left[\frac{kg}{ms}\right]$ is the 2-D convective mass flux density of the VOC
- G is the 2-D domain
- J $\left[\frac{kg}{s}\right]$ is the total extraction rate of the pollutant
- L $\left[\frac{1}{m^2}\right]$ is the 2-D distributed leakage coefficient, $L \geq 0$
- l $\left[\frac{1}{m}\right]$ is the 2-D boundary leakage coefficient
- p $\left[\frac{kg}{ms^2}\right]$ is the air pressure
- p_R $\left[\frac{kg}{ms^2}\right]$ is a reference pressure, $p_R = 100000 Pa$
- $\mathbf{q}$ $\left[\frac{kg}{ms}\right]$ is the 2-D air mass flux density
- R_{eff} $[m]$ is the effective radius
- T $\left[\frac{m^2 s^3}{kg}\right]$ with $0 < \alpha_1 \leq T \leq \alpha_2 < \infty$ is the 2-D transmissivity field in the domain
- T_R $\left[\frac{m^2 s^3}{kg}\right]$ is the reference transmissivity
- u_j $\left[\frac{kg}{s}\right]$, $j = 1, \ldots, n$, are the discharges of the active wells, they are nonnegative real numbers. The wells are assumed to be located at the fixed positions $\mathbf{x}_j, j = 1, \ldots, n$
- u_T $\left[\frac{kg}{s}\right]$ is the total available extraction rate
- V $\left[\frac{kg}{m^2 s}\right]$ is the 2-D mass transfer, i.e., volatilization coefficient

- $V_R = 4 \cdot 10^{-7} \frac{kg}{m^2 s}$ is the reference volatilization coefficient
- $y \left[\frac{kg^2}{m^2 s^4} \right]$ is the square of the air pressure
- $y_R \left[\frac{kg^2}{m^2 s^4} \right]$ is the square of the reference pressure p_R, $y_R = p_R^2$
- $\eta \, [1]$ is the relative mass concentration
- $\eta_S \, [1]$ is a saturation value for the relative pollutant density, $\eta_S = 0.566$
- $\rho \left[\frac{kg}{m^3} \right]$ is the density of the air
- $\rho_R \left[\frac{kg}{m^3} \right]$ is the reference density, $\rho_R = 1.295 \frac{kg}{m^3}$

List of Figures

1.1	NAPL-dissemination in the soil.	2
1.2	Schematic representation of a soil venting installation with a single pumping device.	2
3.1	Schematic representation of a 2-D horizontal domain including active and passive wells.	17
3.2	Comparison of $rY(r,z)$ (solid line) and its quadratic approximation (dashed line) for a well located at $r = 0, z = -2.5$, the water table is at $z = -5$.	28
3.3	Modified Bessel functions K_0, K_1, I_0 and I_1..	29
3.4	Integral radial flow for a 2-D and 3-D point well and a 3-D line well.	30
5.1	Pollutant extraction rate $u\eta(0)$ vs. pumping rate u for a single well with $r_0 = 1$. The dashed line indicates the value π.	46
5.2	Relative concentration η as a function of the distance r for varying initial concentrations at $r = 1.0$ and different values of leakage L and discharge u.	48
5.3	Equilibrium curve $\eta_E(r)$ for $u = 1$ and $L = 100$.	50
5.4	Extraction rate from a single well as function of the discharge u for varying leakage term $L = 0.0, 1.0, \ldots, 10.0$. The initial concentration at $r = 1.0$ is zero.	50
5.5	Relative concentration η as function of the distance r for a bell-shaped volatilization function $V(r)$.	51
5.6	Extraction rate from a single well as function of the discharge u for varying leakage term $L = 1.0, \ldots, 10.0$. The volatilization is assumed to be bell-shaped. The initial concentration at $r = 8.0$ is zero.	52
5.7	Contaminant extraction rate vs. pumping rate u_1.	52
5.8	Streamlines for the optimal pumping rate $u_1 = 0.7$.	53
5.9	Square of pressure and radial flow for a ring sink of radius $r_0 = 0.5$ with constant discharge density and leakage $L = 15$.	55

List of Figures

5.10 Total extraction rates for the pollutant for bell-shaped volatilization, $L = 0.15$ and different values for the discharge. 55

5.11 Total extraction rates for the pollutant for bell-shaped volatilization, $u = 1$ and different values of leakage. 56

5.12 Total extraction rates for the pollutant for bell-shaped volatilization, $L = 15$ and different values for the discharge. 57

6.1 Kertess: Originally layered sandy soils. 61

6.2 Squared differences $[w(x_j) - w(x_j + h)]^2$ of measured data versus spatial distance h between the samples. 76

6.3 Estimator of a covariance. Fitted points are black and irrelevant points are gray colored. 78

6.4 Covariance function $R_1(r)$ for 1-D stationary random field. 83

6.5 Saturation $\theta(h)$ for different layers. 87

6.6 Kertess: Spatial distribution of measurement points. 87

6.7 Kertess: Positions and numeration of wells used for calibration of flow parameters. 88

6.8 Kertess: Discharges for active and passive wells. 89

6.9 Kertess: Transmissivity T $[10^{-10}\frac{m^2 s^3}{kg}]$ after calibration. 91

6.10 Kertess: Leakage term L $[10^{-10}\frac{1}{m^2}]$ after calibration. 92

6.11 Kertess: Pressure y $[atm]$ using calibrated parameters. 93

6.12 Kertess: Flux $\mathbf{q}$ $[\frac{kgm}{s}]$ using calibrated parameters. 93

6.13 Kertess: Positions and numeration of the wells used for calibration of the volatilization coefficient and for the optimization. 94

6.14 Kertess: Original discharges $[\frac{g}{s}]$ of active wells with the total discharge $240 \frac{g}{s}$. 94

6.15 Kertess: Result of kriging – isolines. 95

6.16 Kertess: *Dichlorethen* volatilization coefficient after calibration. 96

6.17 Kertess: *Tetrachlorethen* volatilization coefficient after calibration. 97

6.18 Kertess: *Trichlorethen* volatilization coefficient after calibration. 97

6.19 Kertess: Extraction rates with the original distribution of discharges and the total rate $240 \frac{g}{s}$ for: (A) dichlorethen, (B) tetrachlorethen, (C) trichlorethen and (D) all three NAPL components. 98

7.1 Streamline segment in a finite element. 104

8.1 Kertess: Soil textures and layers above the ground water table of some borehole locations obtained from soil protocols. 112

8.2 Kertess: Pollutant extraction rates. 116

8.3 Kertess: Optimal discharges with the total discharge $240 \frac{g}{s}$ 118

8.4 Kertess: Optimal extraction rates with the total discharge $240 \frac{g}{s}$ 118

8.5 Kertess: Optimal discharges with the total discharge $100 \frac{g}{s}$ 119

8.6	Kertess: Optimal extraction rates with the total discharge 100 $\frac{g}{s}$	119
8.7	Kertess: Optimal discharges with the total discharge 10 $\frac{g}{s}$	120
8.8	Kertess: Optimal extraction rates with the total discharge 10 $\frac{g}{s}$	120
8.9	Kertess: Extraction rates	121
8.10	Kertess: Optimal tetrachlorethen extraction rates with the total discharge 100$\frac{g}{s}$.	122
8.11	Kertess: Optimal tetrachlorethen extraction rates with the total discharge 10$\frac{g}{s}$.	123
8.12	Kertess: Optimization of *tetrachlorethen*	124
8.13	Kertess: Extraction rates for *tetrachlorethen*	124
8.14	Kirchweyhe: Domain and location of active wells.	125
8.15	Kirchweyhe: Volatilization coefficient for Trichlorethen.	126
8.16	Kirchweyhe: Optimal scenario.	126
9.1	Kertess: Discharges Distribution function.	128
9.2	Kertess: Discharges mean function of realizations.	129
9.3	Kertess: Discharges standard deviation function of realizations.	130
9.4	Kertess: Extraction Distribution function.	131
9.5	Kertess: Extraction function of realizations mean.	132
9.6	Kertess: Extraction standard deviation function of realizations.	133
9.7	Kertess: Mean and standard deviation of discharge and Extraction.	135
9.8	Kertess: Mean and standard deviation of extraction rates.	136
9.9	Kertess: Mean and standard deviation of extraction rates.	136

List of Tables

6.1	Kertess: Estimated particle-size distribution.	61
6.2	Estimates of the porosity and bulk density for the originally layered sandy materials and for loose, medium, and dense packings of the particles. The terms d_{10} and d_{60} denote the particle-size at 10% and 60% of the cumulative particle-size distribution function, respectively, and U is the degree of non uniformity.	62
6.3	Estimates of the specific permeability K_w, using Hazen's approximation, the hydraulic conductivity coefficient k_w, and the residual water saturation s_{ob}, of the six sandy soil materials for the originally layered material. The term C^* is an empirical proportionality factor.	63
6.4	Estimates of the specific permeability according to Kozeny-Carman equation for the originally layered material.	64
6.5	Estimates of the Kozeny-Carman coefficients using analytical interpretation of the particle-size distribution function and considering stone and gravel content for six sandy materials of (A) the original layered, (B) the refilled, and (C) the rubble material.	69

List of Tables

6.6 Estimates of bulk density ρ_b, tortuosity coefficient T^*, porosities n^*, n, n', effective hydraulic particle diameter d_w, and specific permeability K_w, according to Kozeny-Carman equation for six sandy materials of the originally layered, refilled, and rubble substrates for (A) the original layered, (B) the refilled, and (C) the rubble material. 71

6.7 Estimated parameters of the pressure-saturation function according to the VGM- $(n, m, m = 1 - 1/n)$ and BC-model (λ) using curve-fitting program RETC for loosely packing of originally layered material with $\theta_r = 0.0, l = 0.5$ and $K_s = 1$ for (A) the original layered sand. 71

6.8 Estimated values of the pressure-saturation function according to the VGM-model $(n, m, m = 1 - 1/n)$ for medium and densely packed originally layered materials assuming θ_s=porosity* (using $\theta_r = 0.0, l = 0.5$ and $K_s = 1$) for (A) the original layered sand. 72

6.9 2-D Covariance Functions and their Fourier Transforms. 85

6.10 Van Genuchten parameters for water $(m = 1 - \frac{1}{n})$. 86

6.11 Chemical and physical properties of some NAPL components. 95

8.1 Kertess: Comparison between estimated (Table 6.6) and measured air permeabilities and porosities for a medium sand. The measured values are obtained using $100 cm^3$ volume soil cores (cf. Lange [Lan92]). 113

8.2 Kertess: Pollutant extraction rates in g/s obtained with the original distribution used by GEO-data compared with the optimized scenarios for total discharge rates of 240, 100, and 10 $\frac{g}{s}$ for sum of three major contaminants and each separately. The values for all components represent the optimized extraction rate considering all the three components simultaneously except for the original distribution where it is the sum of the components above. 116

Chapter 1

Introduction

1.1 Background and Problem

The incorporation of organic solvents, benzene, mineral oil, and other pollutants into the subsurface leads to long-term contaminations of soils and connected ground water systems and may eventually affect whole ecosystems and the air quality. Figure 1.1 (from Einsele et al. [EGS90]) gives a schematic representation of the spreading and fate of halogenated hydrocarbons (HHC) (or in German: "Halogenierte Kohlenwasserstoffe" (HKW)) in both the unsaturated and saturated zones of the soil following a spillage or a leakage. The chlorinated hydrocarbons (CHC) (or in German: "Chlorierte Kohlenwasserstoffe" (CKW)) are a special group of the HKW's. In many cases of soil contaminations with HKW, the polluted soil needs to be replaced completely and cleaned in special facilities at relatively high costs. In cases, where volatile hydrocarbons enter the water-unsaturated zone of the soil, often the *in-situ* technique of "soil venting" is used for remediation.

The soil venting technique is now commonly used for remediation of soil contaminations with volatile organic compounds (VOC) such as, for instance, chlorinated organic hydrocarbon. Soil venting basically means that the soil air is removed from the contaminated soil region by pumping, while the volatile contaminants in the extracted air are collected and removed by special devices above the ground. The pumping of soil air induces a flow of "fresh" air from the atmosphere into and through the soil. Because of the relatively high vapor pressure of the concentration of VOC's in the soil air is gradually increasing by the process of volatilization of the liquid-phase contaminant. The VOC's accumulated in the soil air are eventually transported away from the contaminated soil region together with the mass flow of air. Continuous removal of the dissolved volatile contaminant by the induced mass flow of air leads to a gradual reduction of the hydrocarbon content in the soil. The remediation process is normally regarded as completed when the concentrations of contaminants in the soil drop below a tolerable threshold value limit.

The Figure 1.2 (from Harress et al. [HMS90]) gives a schematic picture of a soil venting installation. In many practical applications a special type of pump or ventilator (in German: "Seitenkanalverdichter") is used which runs at a constant power,

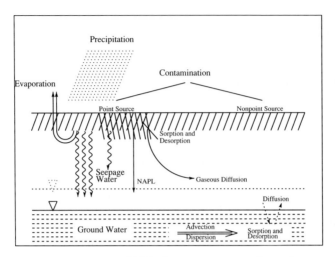

Figure 1.1: NAPL-dissemination in the soil.

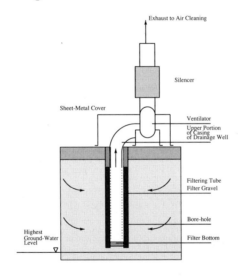

Figure 1.2: Schematic representation of a soil venting installation with a single pumping device.

thus, inducing a stationary air-flow field. The soil-air is removed through extraction (or active) wells, consisting of a filtering tube surrounded by a filtering gravel mantle, which are somehow comparable to wells for pumping ground-water from the saturated zone, however, here installed in the water-unsaturated zone. To improve the efficiency of the installation, additional techniques, such as sealing the surface area surrounding the extraction well, interval pumping, the installation of several extraction wells, and the combination with "passive" wells to increase the infiltration of atmospheric air into the

1.1. Background and Problem

soil at defined places have been proposed and used (e.g., Bruckner and Kugele [BK85], Bruckner et al. [BHH86], Harress et al. [HMS90], Homringhausen and Schwarz [HS92], Rehning [Reh92]).

In general, the soil venting technique is regarded as relatively efficient (Travis and McInnis, [TM92]) because of the more intensive mixing of air in the soil and the normally higher flow velocities of air in the soil compared to ground-water remediation techniques. The efficiency, however, is reduced in case of a long-term soil contamination, in which large amounts of the liquid-phase contaminants (non-aqueous-phase liquid (NAPL)) penetrate more and more immobile pore regions inside of the soil matrix from which it can diffuse only relatively slowly to the mobile pore regions which are in contact to the air phase. Despite of its relative efficiency, the costs of the remediation of soil contaminations, such as non-aqueous-phase liquids (NAPLs), by soil venting can be high. However, the follow-up costs can be substantially higher if remediation is omitted (Mull et al. [MMP92]). Mull et al. [MMP92] also state that potentials for saving costs in soil venting operations are more likely by improving soil survey and operation management than by improving the technical equipment. Since soil venting installations have to be in operation for relatively long time periods the energy consumption can be considerably high and may contribute to environmentally relevant pollution of the air (Quanz and Röhr [QR92]).

Still, it is the common practice to use venting machines having a constant pumping power. This technique, so far, is relatively inefficient because remediation measures often require months or even years while the highest contaminant extraction rates are limited mostly to the initial period of time. Some companies already started to install pumps with controllable air discharge powers assuming that the efficiency of the installation depends largely on the air flow properties of the different soils. The content of contaminants in the extracted soil air is varying considerably with time and is relatively low over long periods of time. Thus, the yield of contaminants, that is the contaminant extraction rate, is often low compared to the effort and the costs of installation, operation (mostly electrical power energy), and for controlling of the venting system facilities.

There is still a significant lack in the theoretical description and modeling of the complex processes involved in soil venting operation for planning and design of soil venting facilities. The main processes occurring during soil air extraction are the transport of contaminants within and together with the soil air, the mass transfers between the fluid and the gas phase of the contaminant, and sorption and dissolution of contaminants. The processes depend on the texture and structure of the soil, the type and distribution of the contaminant in the soil, the temperature, and the actual water content, among other factors. One of the most important parameters is the soil-air permeability which is a function of the porosity and water content, and depends on the spatial distribution of water in the porous medium. Models describing air flow and contaminant transport processes in soils mostly assume a homogeneous porous medium. However, in many cases, the spatial distribution of the subsurface properties is highly heterogeneous. Soils may also be layered and structured, containing cracks, fissures, and macro-pores, or the soil matrix may be formed by aggregates which may lead to the occurrence of preferential flow phenomena. Especially those soils which are contaminated by VOC's have often been strongly anthropogenetically affected and disturbed (buildings, roadsides etc.) which means that its development properties and spatial distributions are highly uncertain and have not been studied as intensively as agricultural or forest soils.

Considering the large number of old-contaminations as well as the new cases the improvement of the efficiency of soil venting operations may be of high significance to reduce costs (see also Kühn [Küh90]).

1.2 Objectives

The objective of the study (on which this book is based) was to develop a model-based technique that may allow to optimize the design of soil venting installations prior to running the operation. It was intended that the optimization technique should be able to consider the case of pollution and site-specific conditions. In mathematical terms, the problem can be denoted as "optimal control of partial differential equations". The overall objective required to develop a simulation model-system containing models to generate model parameters and its spatial distributions, to describe the transport of the contaminants induced by soil venting, and to optimize venting operations. The objective included a computer program and the practical application to typical test cases.

The major idea of the project was

- to apply modern numerical techniques to solve an applied problem of optimizing soil venting operations,

- to combine mathematical, numerical, geo- and pedological knowledge in co-operation considering the practical problems occurring in venting operation, and

- to consider relatively large systems of venting installations comprising several extraction wells.

To achieve the goals, model development was based on data and information from a practical test case operated by the industrial partner.

1.2.1 Mathematical and Numerical Objectives

The specific objective is the mathematical analysis of the system of physical and chemical processes describing the soil venting operation according to real case conditions using data from the Kertess-area as a test case. The first goal is the development of an appropriate mathematical model describing the most important processes, dimensions, and assumptions. The second goal is to develop a numerical model in which the system of coupled nonlinear partial differential equations is treated using the mixed hybrid finite element method. Calibration of the model parameters is intended for using data from the two test cases. Finally, we want to adopt and apply methods for optimal control of partial differential equations. Several test-case scenarios are simulated to demonstrate the ability of the model based optimization tool.

We notice, that each of the mathematical and numerical methods alone is not new. However, new is here the combination and application of the methods to the specific

1.2. Objectives

problem of soil venting, as well as the application of the modeling system for practical determination of optimal strategies of soil venting.

1.2.2 Objectives of the Optimization

The technical problems involved in soil venting are the questions, for instance, how to determine the dimension, the positioning, the number, the installation depth, the tube diameter of the of the extraction wells, and the pumping rates. The pumping rates may, if necessary, also be time variable. The technical objectives are to optimize the costs and efficiency of venting installations, to realistically predict the effective well radii, the required duration of the remediation measure, and the whole remediation process in order to better control the course of the soil remediation, and to optimize the contaminant extraction rates and chemical loads in order to apply the necessary treatment technique to the waste air.

Our intention was, to develop a model system for the optimization of soil venting that is applicable to solve a large number of problems with similar contaminations, and, possibly, to help improving the quality of the cleanup technique itself through intense use of such a model system.

Optimization based on large simulation models has not been possible with inefficient numerical methods. The central scientific aim was to apply the latest numerical methods to improve and to analyze soil remediation using the venting technique. The objective of the BMBF-program was to stimulate the introduction and distribution of advanced mathematical and numerical tools in industry and for engineering purposes to help improving technologies.

Chapter 2

Modeling Soil Venting

2.1 Physical Basis of the Venting Technique

The theory behind the soil venting technique is based on the description of the flow of air in porous media in which the volatile contaminants are being transported.

In principle, the extraction of soil-air from the water-unsaturated zone is a temperature dependent, multidimensional, multicomponent, multi-phase, and – in heterogeneous media – also a two- or multiple pore-region transport problem. The contaminants, e.g., chlorinated hydrocarbons (CHC or CKW), are forming a separate immiscible organic liquid phase (NAPL – Non Aqueous Phase Liquid) in the soil in addition to the solid, the aqueous liquid, and the gas phase.

In the gas phase, the chlorinated hydrocarbon (CKW) is a miscible component of the soil air. CKW may also exist as a dissolved component in the aqueous phase, and adsorbed onto the surface of the solid phase. CKW-contaminants, such as benzene, mostly consist of several different chemical components itself each of which having differing physicochemical properties. Commonly, the specific dense (DNAPL) and specific light (LNAPL) liquid phase organic compounds are distinguished. The movement of these contaminants and its distribution in the soil and ground water systems also differs depending on the pore structure and the soil moisture distribution and content (Mercer and Cohen [MC90]). During soil venting the gas-phase of the chlorinated hydrocarbon is transported out of the soil together with the flow of air. The most critical aspect of the venting technique is how to obtain an optimal relation between the gas transport and volatilization of the fluid phase contaminant CKW in the soil domain affected by an extraction well.

Basic principals of the transport of volatile organic compounds (VOCs) have been studied by Schwille [Sch84], [Sch88] using model experiments. A review about the material properties, models, and possible remediation techniques can be found in Mercer and Cohen [MC90] and Gee et al. [GKLS91]. In case of a non-sealed soil surface, the highly dynamic nature of transport processes affects the movement of organic contaminants in the soil. For instance, NAPL can be transported together with or driven by infiltrating rain water. The movement of water in the soil and the changes of water content as a function of time are modifying the air-filled porosity and the conditions for gas transport in the soil. Soil air extraction may also increase the ground water level. Nahold and Gottheil [NG91], for example, observed an increase of the ground water level of 2m in the soil close to the extraction well. On the other hand, the infiltration of water-vapor-unsaturated atmospheric-air into the ground may contribute to dry out the soil.

The heat and energy balance of the soil is also important for understanding the physicochemical processes during soil venting because of the relatively larger temperature dependency of the density of gases compared to liquids and the drop of temperature during the energy consuming volatilization processes of NAPL and water. Neglecting these temperature dependencies may lead to an underestimation of the remediation time (Lingineni and Dhir [LD92]).

An alternative for increasing the vaporization rates of less volatile organic compounds may be to heat the air that is infiltrating into the soil. An addition of water and water-vapor to the infiltrating air may be used to increase the vaporization by increasing the desorption of NAPL from the solid phase surfaces of relatively dry soils (Donaldson et al. [DMW92]). An induced soil water vapor flow in the soil seemed to increase the volatilization of kerosene (Fine and Yaron [FY93]). Soil venting may also modify the proportions of the chemical fractions of kerosene, i.e., leading to a relative increase in components with larger molecular weight that remain in the soil (Fine and Yaron [FY93]).

The efficiency of soil venting may be reduced by increasing the air flow velocities in the soil (Baehr and Bruell [BB90]). For example, in aggregated soils, the volatilization rate of NAPL may be limited by relatively slow diffusion of the gas-phase NAPL from immobile into the air-mobile pore regions. Especially at rel. high water contents. The assumption of local thermodynamical equilibrium between the gas and liquid phase components may not be valid in such cases.

Of specific importance are also the history of the moisture conditions in the soil during the infiltration of the liquid phase contaminant as well as the duration of the contamination (Travis and McInnis [TM92]). Vapor of organic compounds may be adsorbed onto the soil under conditions of relatively low soil water contents. Culver et al. [CSL91] analyzed the effect of sorption on the transport of LCKW using a two-dimensional numerical model. Under conditions of high soil moisture, the liquid-organic NAPL cannot enter the fine pores of the soil matrix; it remains located predominantly in the larger pore system and so in close contact with the air phase. On the contrary, during the NAPL-infiltration into dry soils the NAPL can also penetrate the meso- and micro-porosity of the soil and may possibly be enclosed during subsequent water infiltration. During long-term contamination, NAPL may gradually be dissolved in the soil water, diffusing into the micropores, and adsorbed onto the solid phase. These long-term contaminations can no longer efficiently be removed by the soil venting technique, since desorption and backward-diffusion through the micropores liquid phase are relatively slow processes and the volatilization rate will become extremely low so that only small amounts of NAPL may evaporate per unit of time.

2.2 Simulation Models

In principle, each case of soil contamination requires the development of a specific model, since all the properties of the contaminants as well as those of the specific soils and geological situations may be different. Some specific models for describing the processes of the soil venting are using one-dimensional finite-difference numerical formulations, cf. Baehr et al. [BHM89b], Croisé et al. [CKS90], Lingineni and Dhir [LD92].

Abriola and Pinder [AP85a] proposed a more general multi-phase model describing flow in two liquid phases (water and NAPL) and in the gas phase of porous media as well as the transport of single components in the phases assuming isothermal conditions. They considered the effect of the compressibility of the matrix and the fluids, the gravity, the composition of the phase components, the mass transfer between the phases, especially the volatilization and dissolution, the capillarity, the diffusion, and the dispersion. The mathematical model consists of a system of three coupled non-linear partial differential equations and two equilibrium conditions. The finite difference method was used for the numerical solution of the spatially one-dimensional case (Abriola and Pinder [AP85b]).

Recently, a comprehensive model (Bear and Nitao [BN92]) has been developed to simulate the remediation of contaminated soils by extraction of soil air which includes also the temperature dependency and non-equilibrium conditions between the phases. Bear and Nitao describe possible discontinuities of a phase, extend the concept of matrix potential to the conditions of multi-phase and multicomponent systems, and include surface effects at dry soil conditions in the retention and permeability function. Bear and Nitao pay special attention to the description of the transport mechanisms by which – in addition to phase pressure gradients (mechanical equilibrium) – also gradients in the chemical potential are involved as driving forces to approach the chemical equilibrium.

Schrefler and Zhan [SX93] propose a model for coupled water and air movement in porous media in which the process of soil consolidation of the solid particles occurs.

Fischer in [Fis95] claim that the gas-water mass transfer coefficient in the kinetics model cannot be expected to be a constant under the condition of soil vapor extraction. Aqueous-gas-phase transfer can be limited by diffusion of the contaminant within the inter-particle water (film).

Further simulation models have been proposed by Corapcioglu and Baehr [CB87], Baehr and Corapcioglu [BC87], Sleep and Sykes [SS89], Falta et al. [FJPW89], Gierke et al. [GHC90], Mendoza and Frind [MF90a], [MF90b]; Gierke et al. [GHM92]; Hornung et al. [HKS96], [HGK$^+$97], Armstrong et al. [AFM94] and Wilson [Wil95].

2.2.1 Soil Gas Transport Models

Basic papers on gas transport in porous media – in the narrow sense – have been published by Cunningham and Williams [CW80], Mason and Malinauskas [MM83], and Thorstenson and Pollock [TP89b], [TP89a] among others. Massmann and Farrier [MF92] describe contaminant-gas transport in porous media including advective air flow driven by alternating barometric air pressure. A most complete description of gas transport in porous media requires to consider – in addition to the molecular diffusion in the multicomponent gas mixture of the soil air and the viscose flow – also the effect of the Knudsen diffusion (Thorstenson and Pollock [TP89b], [TP89a]). Knudsen diffusion describes the effect of the collision of gas molecules with the pore walls of the porous media. The equation that couples the different components of the gas transport process is based on the "Dusty-Gas" model of Mason et al. [MME67], [MM83]. In combination with the Knudsen diffusion, the so-called "Klinkenberg effect" may occur. The Klinkenberg effect describes the increase of the air permeability due to viscose slippering of the gas molecules along the pore walls. This molecular slipping-effect occurs predominantly in fine pores and at decreasing pressure in the air phase. The total gas transport in soils is according to Massmann and Farrier [MF92] a combination of the diffusive flux, induced by gradient of the gas concentration or partial pressures (Knudsen diffusion, molecular diffusion, and transition-region diffusion), the pressure dependent flow (viscose flow and slipping along surfaces), and the combined flow depending on the total and partial gas pressure gradient. Because of the Knudsen diffusion, also non-equimolar flow may occur. A model that includes all the above mentioned processes in a multicomponent gas mixture is described by Thorstenson and Pollock [TP89b].

Other models consider the influence of gravity on the transport of specific dense organic vapors and soils with relatively high permeability (Falta et al. [FJPW89]), (Mendoza and Frind [MF90a], [MF90b]). The so-called Stefan-Maxwell-equation describes the diffusive portion of the gas flow of a multicomponent gas mixture. Based on analysis of comparative simulations using complex gas flux descriptions Massmann and Farrier [MF92] found that the classical convection-dispersion equation may be used to describe gas flows in unsaturated soils provided that the permeabilities of the porous medium are larger than 10^{-10} cm^2. In less permeable soils the effect of the Knudsen diffusion on the flux increases such that the convection-dispersion equation may overestimate the flow rates. The Stefan-Maxwell equation on the other hand may underestimate the fluxes in porous media of relatively high permeability and may overestimate the flux considerably in media with relatively low permeability.

2.2.2 Phase Mass Transfer Models

Organic contaminants may occur in the soils in form of dissolved components either in the liquid NAPL or aqueous phase. Assuming local thermodynamic equilibrium, most models are using Henry's law to determine the distribution of the compounds between the aqueous and the air phase and Raoult's law to determine the distribution between NAPL and air itself (Mercer and Cohen [MC90]). The temperature dependency of the vapor pressure is described by the Clausius-Clapeyron equation (Corapcioglu and Baehr [CB87]) or by the vapor-pressure relation of Antoine (Lingineni and Dhir [LD92]). Evaporation may be regarded as the basic cause of the vapor transport in soils. Advective flow driven by pressure- and density-gradients may play an important role for the successive spreading or distribution of contaminants in the subsoil and may contribute to the contamination of ground water. The assumption of local thermodynamic equilibrium is used also for other constitutive relations, e.g., to interrelate the mass fractions

2.2. Simulation Models

of the remaining species of all phases (Falta et al. [FJPW89], Sleep and Sykes [SS89], Mendoza and Frind [MF90a].

2.2.3 Multi-phase Permeability Models

Basic parameters of multi-phase transport models are the air, water, and NAPL permeability and the saturation-pressure-relations. Lenhard and Parker [LP87a] extended approaches to describe the model parameters of two-phase to three-phase systems. Lenhard and Parker [LP87b] and Parker and Lenhard [PL87] further extended the concept by including the hysteresis of the pressure-saturation relations. Lenhard et al. [LPK91] could demonstrate through non-stationary air-flow experiments that the consideration of hysteresis in a numerical model of multi-phase flow system improved the results considerably, however, by the cost of adding another model parameter.

Shan et al. [SFJ92] developed an analytical solution to describe soil air extraction assuming stationary flow in order to determine the horizontal and vertical gas conductivity. The model of Shan et al. [SFJ92] can be used to determine the permeability for gas in soils using data of pumping experiments. The model can also be utilized for relatively simple optimizations of extraction wells. Another analytical approach was proposed by Baehr and Hult [BH91] to solve the two-dimensional cylindrical gas flow equation combined with an numerical optimization scheme to calculate gas permeabilities from pumping experiments. The method, in analogy to the inverse problem, is commonly used in the well-hydraulic literature. Demond and Roberts [DR91] found that well-known models for estimating permeability functions can be of limited use when applied to organic fluids, since they regularly overestimate the permeability. In addition, they found a hysteretic behavior of the function for organic liquids.

2.2.4 Biological Decay

Biological decay of the organic vapor phase is strongly dependent on the conditions and often relatively slow and by some authors regarded as a minor process with respect to soil remediation (Braun et al. [BPB93], Coffa et al. [CBUS92]). Ostendorf and Kampbell [OK91], on the other hand, state that microbial decay of hydrocarbon vapors is an important process. To accelerate the decay of organic compounds in the soil systems, special techniques, such as ozonization, have been analyzed (e.g., Mercer and Cohen [MC90], Harress [HMS90]). Chen et al. [CAA$^+$92] studied the transport and biological decay of benzene and toluene in sandy aquifers using steady-state flow soil-column experiments and a comprehensive simulation model of advective and dispersive transport, mass transfer between the solid, fluid, and gas phases and the biomass as well as kinetically controlled decay.

Schumacher and Slodička [SS96] have theoretically studied the influence of several physical parameters on the spreading of the microbes. Here the pollutant is considered as a nutrient for the microbes in low concentrations while being poisonous in higher concentrations.

2.2.5 Measurement Configuration

The soil conditions at the beginning of a remediation process and the configuring the technical equipment of soil venting operations are mostly determined through measurements of the contaminant concentration in the soil air at different locations and in several soil depths (Grathwohl [Gra90]). Based on air pumping tests, chemical analyzes of extracted soil air, and the knowledge of the geology and soil conditions (e.g., Homringhausen and Schwarz [HS92]), the amount and distribution of the contaminants in the soil is estimated experimentally. An important parameter herein is the residual saturation, which is the amount of the contaminant, that remains in the soil pore system after the liquid phase contaminant from a spillage has passed through the soil (e.g., Schwille [Sch84]). Baehr et al. [BHM89b] studied residual saturations of benzen in homogeneous wet sands, Pantazidou and Sitar [PS93] did experimental studies of the movement and spreading patterns of LNAPL (Kerosene) in the vadose zone and the connected displacement of soil pore water by the contaminant.

2.2.6 Heterogeneous Media

The spreading of differently dense NAPL has been studied by Kueper and Frind [KF91a], [KF91b], however, for water saturated porous media using two-dimensional model simulations of two-phase flow in heterogeneous porous media and measured data from a sandy aquifer. Kueper and Frind found, that the lateral spread-out of contaminants increases with increasing density and heterogeneity of the medium. The spreading of the contaminants was also studied by Poulsen and Kueper [PK92] by excavating soil profiles in the field. They found highly heterogeneous distributions of contaminants depending on the soil structure, and they questioned if the total distribution of contaminants may be deducted sufficiently exact by results of point measurements of the contaminant content.

2.2.7 Structured Soil

An additional complexity with respect to the contaminant transport can be expected in structured soils. Such soils cannot be regarded as homogeneous porous media. They contain cracks and fissures or may be aggregated. Soil air may flow in structured soils along preferential pathways much faster than expected and by-passing the less permeable soil matrix or porous blocks. Thus, among other things, thermodynamic equilibrium relations between the concentration in the fluid phase and the partial pressure of a component in the gas phase cannot be assumed a priori (Gierke et al. [GHM92]). For the description of the mass transfer caused by intra-aggregate diffusion hardly any estimates for the transfer coefficients of the kinetic first-order rate equations are available (Sleep and Sykes [SS89]). For mobile-immobile type pore systems, Brusseau [Bru91] described a model of stationary gas transport in aggregated heterogeneous soils in analogy to the two-region model of solute transport with rate-limited sorption. Powers et al. [PLAWJ91] found deviations from the local dissolution-equilibrium in ground water,

2.2. Simulation Models

and they supposed the mass transfer across a boundary layer between NAPL and water is the limiting factor for the dissolution of NAPL in ground water.

2.2.8 Mathematical Aspects

Flow in porous media can be described, according to a mass balance law, by a nonlinear parabolic partial differential equation (or by a system in more general case). The idea of studying abstract nonlinear parabolic PDEs by means of the theory of nonlinear semigroups of contractions in Banach spaces was proposed by Brezis [Bre70]. This approach has not only theoretical but also numerical aspects. It suggests some algorithms for time-discretization in order to approximate the exact solution. Some algorithms are based on Crandall-Liggett formula (cf. Crandall Liggett [CL71]), others on the nonlinear Chernoff formula (see Berger, Brezis and Rogers [BBR79], Magenes, Nochetto and Verdi [MNV87]). Another approach based on a linearization was proposed by Jäger and Kačur [JK91].

A large number of applied papers (without numerical discussions) are suggesting different numerical schemes for computing transport in porous media, e.g., Abriola and Pinder [AP85b], Culver et al. [CSL91], Forsyth and Shao [FS91], Falta et al. [FPJW92], Celia and Bining[CB92], Zeitoun and Pinder [ZP93], Adenekan, Patzek and Pruess [APP93], Schrefler and Zhan [SX93].

All the different approaches for space discretization (finite differences, finite elements, finite volumes) have been used in the literature for models of soil venting. This variability of methods has also its historical reason. From the physical point of view, the approaches with the most exact mass balance should be preferred (mixed FEM, finite volumes). Finite elements became one of the most applied methods for solving the transport in porous media (cf. Ciarlet and Lions [CL91]). The main reasons for this are

(i) FEMs are based on the weak variational formulation of boundary and initial value problems,

(ii) FEM can be applied to domains of arbitrary shape,

(iii) FEMs lead by their nature to unstructured meshes,

(iv) robustness of the method,

(v) mathematical foundations.

The optimization-based approaches have been often applied in water remediation (cf. Kuo, Michel and Gray [KMW92], Gorelick [Gor90]), but the applications in soil venting are rare. Most of the nonlinear optimization algorithms need the derivatives of the cost functional (cf. Lions [Lio71]) – which is not available in real cases. We will use and describe a simple method for optimization without the derivatives of the cost functional (see Section 7.4).

2.2.9 Remarks on Preferential Gas Flow in Structured Soils

Preferential gas transport in structured soils may occur during soil venting in case a large part of the air flow takes place in a relatively limited volume of the soil pore system through inter-aggregate pores, soil macro-pores, fissures and cracks.

In soils where relatively small and homogeneously distributed aggregates are forming a rel. homogeneous macro-pore system we may assume that the macro-pore system will be flown through evenly. For such soils approaches such as the two-region model (e.g., Brusseau [Bru91], Gierke et al. [GHM92] may be used for a stationary solute transport in soils with mobile and immobile pore regions conceptionelly applied to air flow. The mass transfer of gas between the soil matrix can be described by approaches using first-order type differential equations. In media with a dual-porosity system also two-scale, two-region models (e.g., Hornung [Hor92]) may be used that allow also instationary flow conditions. In the two-scale approach, only the flow and transport in macro-pore system is described macroscopically. The mass transfer, as a sink term in the macroscopic equation, can be approximated with models of diffusive and convective transport in aggregates. On the other hand, dual-permeability models (e.g., Gerke and van Genuchten [GvG93]) may not be useful for the above described case of air flow, because the assumption of a continuous gas phase in the matrix pore system seems to be unlikely except for relatively dry soils.

Gimmi in [Gim94] has studied the gaseous mobile-immobile-type transport equations in which an "effective" diffusion coefficient D_e can be derived. D_e was found to be dependent on time and distance, i.e., for relatively slow transport in the macro-pores, it approximates rapidly the constant equilibrium diffusion coefficient. The definition of boundary condition in the case without pumping is a problem, because of the back-diffusion process through the soil surface. Sorption and hydratation effects have also been studied.

The situation may be different for soils interspersed by few continuous cracks and fissures. In these soils a bypass or shortcut air flow may occur through the cracks, and the surrounding soil matrix may still have a continuous gas phase and is participating in the total flow of air. Here, dual permeability models would be more appropriate or such models which describe flow in discrete cracks. The degree of preferential air flow in cracks will be influenced strongly by the pressure gradients and flow velocities that are induced in the crack or fissure system by soil venting.

The development of a gas flow of the type described above is depending on the pressure gradient induced from outside. On the contrary to water flow, the effect of the gravity on gas flow is mostly very limited except for specifically dense organic vapor components. The gravity effect can be modeled as potentiometric gas pressure (Thorstenson and Pollock [TP89a]). Models for soils with relatively high permeabilities and specifically dense vapor (DNAPL) have been proposed by Falta et al. [FJPW89] and Mendoza and Frind [MF90a], [MF90b]. Unclear are the effects of preferential gas flow on water movement and vice versa. The preferential water flow in macro-pores may induce a displacement of air into the surrounding pores or into the atmosphere.

Chapter 3

Stationary Problem and Optimal Control

Soil venting is an in-situ process to remove the volatile organic compounds from the unsaturated zone. The extraction well creates a negative differential pressure which generates a gas flow towards the probe. After removal the VOCs existing in the equilibrium conditions, VOC will enter the gaseous phase from the liquid phase due the phase change. Hence the discharges of the active probes must be relatively small, in the other case the concentration of VOCs in the extracted gas will be low. Differential pressures applied to the extraction well typically range from $15''$–$350''$ of water (cf. Hiller [Hil]). The influence of a vapor extraction system depends on many parameters (stratigraphy, soil type, length and position of the probes, discharge, VOC distribution and its concentration in the soil matrix, soil surface, depth of the vadose zone). In the case when the soil surface is not covered, the range of influence of a single well is relatively small. In order to increase the radius of influence one can use full or partially cover of the soil surface. According to the field tests, the time required to develop the steady state flow and the effective radius ranges between 15–30 minutes (cf. Hiller [Hil]). On the other hand the movement of the air inside the porous matrix is slow (it can take days to arrive from the range of the remediation site into the active probe) and the whole soil venting process takes place over a few years. From this point of view it is natural to study the steady state problem first.

Throughout this book we will consider a typical situation, where all the geological layers are horizontally located, the soil surface is sealed (e.g., by beton, asphalt, iron plate), the remediation site is insulated from the bottom by an impervious layer or water table, and the vertical depth of the well screen is almost equal to the thickness of the vadose zone. Then the air flow field inside the unsaturated zone is essentially horizontal. Now, one can integrate the gas flow between the soil surface and the bottom and in this way to reduce the 3-D problem to the vertically integrated 2-D problem.

In this chapter we state the main problems of air flow, contaminant transport and the optimization. For more detailed mathematical description see Chapter 4. We make the following assumptions.

Major Assumptions

1. The gas flow is stationary.
2. The gas obeys the ideal gas law.
3. The polluted domain is not perfectly insulated, but its upper boundary and the lateral boundary are leaky.
4. The under-pressure at the active wells is small compared with the absolute value of the atmospheric pressure.
5. The concentration of the VOCs in the gaseous phase is small compared with the total gas density; hence the gas flow can be modeled without explicitly taking the contaminant into account.
6. The volatilization, i.e., the phase change of the VOCs from fluid to gaseous, can be described by a linear relationship. The transfer coefficient for this phase change does not depend on the Darcy velocity of the gas, but only on the soil properties and the concentration of the VOCs in the fluid phase.
7. The diffusion of the VOCs is small compared with the convection, i.e., the Péclet number is large.
8. The concentration of the VOCs in fluid phase is so large that its decrease during the modeling period is negligible.
9. The phreatic surface is not significantly influenced by the pumping procedure.

3.1 Basic and State Equations, Objective Function

In this section we give a model description of stationary pollutant transport induced by a stationary gas flow in a 2-D leaky domain with N small circular holes (these represent the wells).

Let G be an open domain in $\mathbb{R}^2$. Let $\mathbf{x}_j \in G$ ($j = 1, \ldots, N$) be a set of interior points and $R_j > 0$ radii such that the closure of the disks $G_j = \{\mathbf{x} \in \mathbb{R}^2 : |\mathbf{x}-\mathbf{x}_j| < R_j\}$ belongs to G, i.e., $\overline{G_j} \subset G$. We assume that the closed disks do not intersect : $\overline{G_i} \cap \overline{G_j} = \emptyset$ for $i \neq j$. We denote $\Gamma_j = \partial G_j$ and $G_H = G \setminus \bigcup_{j=1}^{N} \overline{G_j}$ and $\Gamma = \partial G$, see Figure 3.1.

For $j = 1, \ldots, n$ ($n < N$) the holes act as active wells for the gas flow (and the pollutant flow as well). The active holes and their boundary are denoted by

$$G_A = \bigcup_{j=1}^{n} G_j, \quad \Gamma_A = \partial G_A = \bigcup_{j=1}^{n} \Gamma_j.$$

For $j = n+1, \ldots, N$ the passive wells are holes with prescribed values for the square of the gas pressure. There is no pollutant inflow through these holes. The passive holes and their boundary are denoted by

$$G_P = \bigcup_{j=n+1}^{N} G_j, \quad \Gamma_P = \partial G_P = \bigcup_{j=n+1}^{N} \Gamma_j.$$

The vector $\boldsymbol{\nu}$ stands for the exterior normal (with respect to G_H) on Γ or Γ_j, resp.

3.1. Basic and State Equations, Objective Function

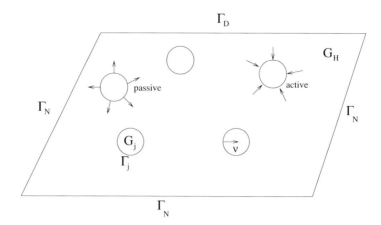

Figure 3.1: Schematic representation of a 2-D horizontal domain including active and passive wells.

3.1.1 Air Flow

First we are going to model the gas flow. One of the most discussed things by modeling is the problem how to describe wells. We state some of the typical boundary conditions used for active or passive air probes. Let us point out that the boundary conditions used for water or oil pumping wells could be different (cf. Slodička [Slo98]).

Data: The transmissivity and the leakage term are given as space-dependent functions $T, L : G \to \mathbb{R}$. The quantity y_R is a given reference value. The functions y_D, q_N, and y_P are the boundary values on Γ_D, Γ_N, and Γ_P, resp., where y_P is a given constant, for example $y_P = p_R^2$, where p_R is the atmospheric pressure. The function $l : \Gamma_N \to \mathbb{R}$ is a prescribed leakage function for the boundary. Here the control variable $u = (u_1, \ldots, u_n) \in \mathbb{R}^n$ is assumed to be given.

Unknown: The function $y : G_H \to \mathbb{R}$ has to be found and it describes the square of the pressure $y = p^2$; the gas density ρ can be calculated from the ideal gas law $\rho = \text{const } p$.

Notations: The vector field $\mathbf{q} = -T(\mathbf{x})\nabla y$ is the air mass flow.

We now consider the following **boundary value problem (BVP)** in G_H.

$$\begin{cases} \nabla \cdot \mathbf{q} &= LT(y_R - y) & \text{in } G_H \\ \mathbf{q} &= -T\nabla y & \text{in } G_H \\ \mathbf{q} \cdot \boldsymbol{\nu} &= -lT(y_R - y) + q_N & \text{on } \Gamma_N \\ y &= y_D & \text{on } \Gamma_D \\ y &= y_P & \text{on } \Gamma_P \end{cases} \quad (3.1)$$

together with one of the following boundary conditions for the active wells:

(P) **Pressure Condition** The u_j are understood as under-pressures at the active wells, i.e.,

$$y = y_R - u_j \text{ on each } \Gamma_j \subset \Gamma_A \quad (3.2)$$

(F) **Flux Condition** The u_j are understood as discharges of the wells, and the normal flux is uniformly distributed along the boundary of the well, i.e.,

$$\mathbf{q} \cdot \boldsymbol{\nu} = -u_j \frac{T(x)}{\int_{\Gamma_j} T(x) \, d\Gamma} \quad \text{on each } \Gamma_j \subset \Gamma_A \tag{3.3}$$

(D) **Discharge Condition** The u_j are understood as discharges of the wells, and it is assumed that a constant under-pressure builds up on the boundary such that the prescribed discharge is obtained, i.e., there are numbers $\gamma_j \in \mathbb{R}$ such that

$$y = \gamma_j \text{ on each } \Gamma_j \subset \Gamma_A \text{ and } \int_{\Gamma_j} \mathbf{q} \cdot \boldsymbol{\nu} \, d\Gamma = u_j. \tag{3.4}$$

Remark 3.1 With the choice (F) for the active wells, the extraction rate condition

$$\int_{\Gamma_j} \mathbf{q} \cdot \boldsymbol{\nu} \, d\Gamma = u_j$$

is automatically satisfied. Boundary condition (P) and (F) have the advantage that the dependence on the control variables u_j is explicitly given, while (D) seems to be the most natural choice. □

Remark 3.2 Certain SVE devices are designed such that the active wells simultaneously operate in groups. This means that all wells belonging to one group have the same under-pressures. In this case the number of control variables u_k reduces to the number m of groups ($m < n$). Let $W_k \subset \{1, \ldots, n\}$ denote the wells belonging to the k-th group ($k = 1, \ldots, m$). In the case of the pressure condition [P] the boundary condition then becomes

$$y = y_R - u_k \text{ on each } \Gamma_j, \ j \in W_k, \ k = 1, \ldots, m.$$

In the case of the discharge condition [D] the resulting boundary condition is that there are numbers $\gamma_k \in \mathbb{R}$ such that

$$y = \gamma_k \text{ on each } \Gamma_j, \ j \in W_k \text{ and } \sum_{j \in W_k} \int_{\Gamma_j} \mathbf{q} \cdot \boldsymbol{\nu} \, d\Gamma = u_k, \ k = 1, \ldots, m. \tag{3.5}$$

□

In the following we assume that the choice of the data is such that $y > 0$ in G_H so that p and ρ are defined.

3.1.2 Contaminant Transport

Now we are modeling the contaminant transport.

Data: From the boundary value problem of the previous section we obtain the air density ρ and the air mass flow $\mathbf{q}$.

There is a distributed volatilization source described by a space-dependent function $V(\mathbf{x}) \geq 0$.

In addition, $D \geq 0$ is the diffusion coefficient, C_D are Dirichlet boundary values for C, $\eta_S > 0$ is a saturation value for the relative pollutant density, and ρ_R is the density of the air corresponding to the reference value p_R.

Unknown: The relative mass concentration η in the domain G_H has to be found.

Notations: $C = \rho \eta d$ with the vertical thickness d of the domain is the mass density of the contaminant, $\mathbf{f} = \eta \mathbf{q}$ is the convective mass flux, and $\mathbf{F} = -D\rho_R \nabla \eta + \mathbf{f}$ is the total mass flux of the pollutant.

For the formulation of the boundary conditions we have to split the outer boundary Γ into two parts $\Gamma_+ = \{\mathbf{x} \in \Gamma : \mathbf{q} \cdot \boldsymbol{\nu} \geq 0\}$ and $\Gamma_- = \{\mathbf{x} \in \Gamma : \mathbf{q} \cdot \boldsymbol{\nu} < 0\}$.

The **boundary value problem** in G_H is the following

$$\begin{cases} \nabla \cdot \mathbf{F} = V(\eta_S - \eta) & \text{in } G_H \\ \mathbf{f} = \eta \mathbf{q} & \text{in } G_H \\ \mathbf{F} = -D\rho_R \nabla \eta + \mathbf{f} & \text{in } G_H \\ \mathbf{F} \cdot \boldsymbol{\nu} = \mathbf{f} \cdot \boldsymbol{\nu} & \text{on } \Gamma_A \cup \Gamma_+ \\ \eta = 0 & \text{on } \Gamma_P \cup \Gamma_-. \end{cases} \qquad (3.6)$$

For $D = 0$, the boundary conditions for the outer boundary Γ and the active wells have to be dropped.

3.1.3 Optimal Control

We are now in the position to describe the optimal control problem.

Data: We assume that for a given set of control variables $u_1, \ldots, u_n$ the two BVPs from the two previous sections can be solved. In this way we get the mass flux $\mathbf{F}$.

We further assume that a certain set $U \subset \mathbb{R}^n_+$ of admissible controls is prescribed. An example for this is

$$U = \{u = (u_1, \ldots, u_n) : 0 \leq u_j \leq u_j^+ \text{ for } j = 1, \ldots, n, \text{ and } \sum_{j=1}^n u_j \leq u_T\}$$

where $u_1^+, \ldots, u_n^+$ and $u_T > 0$ are prescribed upper limits.

In addition, functions f_j, $g_j : \mathbb{R} \to \mathbb{R}_+$ are assumed to be given.

Unknown: What has to be found is the optimal choice of the control variables $u_1, \ldots, u_n$.

Notations: We consider a general cost functional of the form

$$J(u_1, \ldots, u_n) = -\sum_{j=1}^n f_j \left(\int_{\Gamma_j} \mathbf{F} \cdot \boldsymbol{\nu} \, d\Gamma \right) + \sum_{j=1}^n g_j(u_j) \qquad (3.7)$$

The first sum in the cost functional is a measure for the extraction of pollutant from the soil, while the second sum measures the costs of generating the gas flow field. A typical example for a cost functional is given by $f_j(z) = c_1 z$ and $g_j(z) = c_2 z$ with constants $c_1, c_2 \geq 0$; then we get

$$J(u) = -c_1 \sum_{j=1}^{n} \int_{\Gamma_j} \mathbf{F} \cdot \boldsymbol{\nu} \, d\Gamma + c_2 \sum_{j=1}^{n} u_j = -c_1 \int_{\Gamma_A} \mathbf{F} \cdot \boldsymbol{\nu} \, d\Gamma + c_2 \sum_{j=1}^{n} u_j. \tag{3.8}$$

In the special case $c_1 = 1$, $c_2 = 0$ we get

$$J(u) = -\int_{\Gamma_A} \mathbf{F} \cdot \boldsymbol{\nu} \, d\Gamma$$

The **optimal control problem**

$$\inf \{J(u) : u \in U\} \tag{3.9}$$

is to find $u = (u_1, \ldots, u_n) \in U$ such that the cost functional is minimal among all $u \in U$, where the set $U \subset \mathbb{R}_+^n$ is the set of admissible controls.

Remark 3.3 For the optimization of the well locations see Section 10.2. □

Remark 3.4 Let us note that from the physical point of view the discharges of active wells must be relatively small. In the opposite case one can obtain a negative pressure at the extraction wells, which is not realistic. □

3.2 Mathematical Analysis, Simplifications and Modeling Aspects

Sometimes it is very useful to study some simple cases and/or to re-scale the original problem into to a "dimensionless" form in order to obtain a better feeling about the solution. We show the reader a few simplified formulations. In the case of a not perfectly insulated soil surface we study how to describe the "leaky surface" in 3-D using the leakage term L in 2-D case. Further we define an air flow problem for the point sinks, i.e., when the active wells are considered as points. At the end of this section we briefly outline the idea of the streamline method.

3.2.1 Dimensionless Form

Constant Coefficients

We assume that L, T, and V are constants and study the following system of equations.
For the gas flow

$$\begin{cases} \nabla \cdot \mathbf{q} = LT(y_R - y) & \text{in } G_H \\ \mathbf{q} = -T\nabla y & \text{in } G_H. \end{cases} \tag{3.10}$$

3.2. Mathematical Analysis, Simplifications ...

For the contaminant transport

$$\begin{cases} \nabla \cdot \mathbf{F} = V(\eta_S - \eta) & \text{in } G_H \\ \mathbf{F} = \eta \mathbf{q} & \text{in } G_H. \end{cases} \quad (3.11)$$

The objective function

$$J(u) = \int_{\Gamma_A} \mathbf{F} \cdot \boldsymbol{\nu} \, d\Gamma. \quad (3.12)$$

Now let ε be a characteristic length of the problem. Then we use the following scaling of the variables (here the hat does not indicate the Fourier transform, but rescaling):

For the gas flow

$$\boxed{\mathbf{x} = \varepsilon \hat{\mathbf{x}}, \quad \mathbf{q} = \varepsilon V \hat{\mathbf{q}}, \quad L = \frac{\hat{L}}{\varepsilon^2}, \quad y = y_R + \frac{\varepsilon^2 V}{T} \hat{y}}.$$

For the contaminant transport

$$\boxed{\mathbf{F} = \varepsilon \eta_S V \hat{\mathbf{F}}, \quad \eta = \eta_S \hat{\eta}}.$$

For the control and objective function

$$\boxed{u = \varepsilon^2 V \hat{u}, \quad J(u) = \varepsilon^2 \eta_S V \hat{J}(\hat{u})}.$$

Since $\nabla = \frac{1}{\varepsilon} \hat{\nabla}$, we get from this

$$\begin{cases} \hat{\nabla} \cdot \hat{\mathbf{q}} = -\hat{L} \hat{y} & \text{in } \hat{G}_H \\ \hat{\mathbf{q}} = -\hat{\nabla} \hat{y} & \text{in } \hat{G}_H \end{cases} \quad (3.13)$$

$$\begin{cases} \hat{\nabla} \cdot \hat{\mathbf{F}} = 1 - \hat{\eta} & \text{in } \hat{G}_H \\ \hat{\mathbf{F}} = \hat{\eta} \hat{\mathbf{q}} & \text{in } \hat{G}_H \end{cases} \quad (3.14)$$

$$\hat{J}(u) = \int_{\hat{\Gamma}_A} \hat{\mathbf{F}} \cdot \boldsymbol{\nu} \, d\hat{\Gamma}. \quad (3.15)$$

Remark 3.5 If we write $\varepsilon = R_{\text{eff}} \, \hat{\varepsilon}$, then we get from formula (3.44)

$$\hat{L} = z_{10} \hat{\varepsilon}^2. \quad (3.16)$$

□

Dirac Sinks

If we deal with Dirac-type sinks, then we have to use the following scaling law.

Lemma 3.6 Let f be a regular distribution on $\mathbb{R}^d$. Then after substituting $\mathbf{x} = \varepsilon \hat{\mathbf{x}}$ one has for the scaled distribution
$$\hat{f}(\hat{\mathbf{x}}) = \varepsilon^d f(\mathbf{x}).$$

Proof: For any test function φ one has
$$\langle \hat{f}, \varphi \rangle = \int_{\mathbb{R}^d} f(\mathbf{x})\varphi(\mathbf{x})\, d\mathbf{x} = \varepsilon^d \int_{\mathbb{R}^d} f(\hat{\mathbf{x}})\varphi(\hat{\mathbf{x}})\, d\hat{\mathbf{x}} = \langle \varepsilon^d f, \varphi \rangle.$$ □

If we start from
$$\nabla \cdot \mathbf{q} = LT(y_R - y) - \sum_j u_j \delta(\mathbf{x} - \mathbf{x}_j)$$

and use
$$u_j = \varepsilon^2 V \hat{u}_j$$

then we get
$$\hat{\nabla} \cdot \hat{\mathbf{q}} = -\hat{L}\hat{y} - \sum_j \hat{u}_j \delta(\hat{\mathbf{x}} - \hat{\mathbf{x}}_j).$$

In the same way, starting from
$$\nabla \cdot \mathbf{F} = V(\eta_S - \eta) - \sum_j \eta v_j \delta(\mathbf{x} - \mathbf{x}_j)$$

we get
$$\nabla \cdot \hat{\mathbf{F}} = 1 - \hat{\eta} - \sum_j \hat{\eta}\hat{u}_j \delta(\hat{\mathbf{x}} - \hat{\mathbf{x}}_j).$$

If the objective function is
$$J(u) = \sum_j u_j \eta(\mathbf{x}_j)$$

then we get
$$\hat{J}(\hat{u}) = \sum_j \hat{u}_j \hat{\eta}(\hat{\mathbf{x}}_j).$$

Variable Coefficients

In the case of variable coefficients L, T, and V one proceeds in the following way. First one defines reference values T_R and V_R. Then one uses the following scaling.

For the gas flow

$$\boxed{\mathbf{x} = \varepsilon \hat{\mathbf{x}}, \quad \mathbf{q} = \varepsilon V_R \hat{\mathbf{q}}, \quad L(\mathbf{x}) = \frac{\hat{L}(\mathbf{x})}{\varepsilon^2}, \quad T(\mathbf{x}) = T_R \hat{T}(\mathbf{x}), \quad y = y_R + \frac{\varepsilon^2 V_R}{T_R} \hat{y}}$$

3.2. Mathematical Analysis, Simplifications ...

For the contaminant transport

$$\boxed{\mathbf{F} = \varepsilon \eta_S V_R \hat{\mathbf{F}}, \quad \eta = \eta_S \hat{\eta}, \quad V(\mathbf{x}) = V_R \hat{V}(\mathbf{x})}.$$

For the control and the objective function

$$\boxed{u = \varepsilon^2 V_R \hat{u}, \quad J(u) = \varepsilon^2 \eta_S V_R \hat{J}(\hat{u})}.$$

Then one gets

$$\begin{cases} \hat{\nabla} \cdot \hat{\mathbf{q}} = -\hat{L}(\mathbf{x})\hat{T}(\mathbf{x})\hat{y} & \text{in } \hat{G}_H \\ \hat{\mathbf{q}} = -\hat{T}(\mathbf{x})\hat{\nabla}\hat{y} & \text{in } \hat{G}_H \end{cases} \tag{3.17}$$

$$\begin{cases} \hat{\nabla} \cdot \hat{\mathbf{F}} = \hat{V}(\mathbf{x})(1-\hat{\eta}) & \text{in } \hat{G}_H \\ \hat{\mathbf{F}} = \hat{\eta}\hat{\mathbf{q}} & \text{in } \hat{G}_H \end{cases} \tag{3.18}$$

$$\hat{J}(\hat{u}) = \int_{\hat{\Gamma}_A} \hat{\mathbf{F}} \cdot \boldsymbol{\nu} \, d\hat{\Gamma}. \tag{3.19}$$

Dimensionless Boundary Value Problems

For the formulation of the full boundary value problems we also need to introduce the following dimensionless quantities

$$\boxed{l(\mathbf{x}) = \frac{\hat{l}(\mathbf{x})}{\varepsilon}, \quad \mathbf{q}_N = \varepsilon V_R \hat{\mathbf{q}}_N, \quad y_D = y_R + \frac{\varepsilon^2 V_R}{T_R}\hat{y}_D, \quad y_P = y_R + \frac{\varepsilon^2 V_R}{T_R}\hat{y}_P}.$$

Then we get in the case of discharge boundary conditions [D] the following problem for the gas flow

$$\begin{cases} \hat{\nabla} \cdot \hat{\mathbf{q}} &= -\hat{L}(\hat{\mathbf{x}})\hat{T}(\hat{\mathbf{x}})\hat{y} & \text{in } \hat{G}_H \\ \hat{\mathbf{q}} &= -\hat{T}(\hat{\mathbf{x}})\hat{\nabla}\hat{y} & \text{in } \hat{G}_H \\ \hat{\mathbf{q}} \cdot \boldsymbol{\nu} &= \hat{l}(\hat{\mathbf{x}})\hat{T}(\hat{\mathbf{x}})\hat{y} + \hat{q}_N & \text{on } \hat{\Gamma}_N \\ \hat{y} &= \hat{y}_D & \text{on } \hat{\Gamma}_D \\ \hat{y} &= \hat{y}_P & \text{on } \hat{\Gamma}_P \\ \hat{y} &= \hat{\gamma}_j & \text{on each } \hat{\Gamma}_j \subset \hat{\Gamma}_A \\ \int_{\hat{\Gamma}_j} \hat{\mathbf{q}} \cdot \boldsymbol{\nu} \, d\hat{\Gamma} &= \hat{u}_j & \text{for each } \hat{\Gamma}_j \subset \hat{\Gamma}_A. \end{cases} \tag{3.20}$$

Dropping diffusion we obtain the following problem for the contaminant transport

$$\begin{cases} \hat{\nabla} \cdot \hat{\mathbf{F}} = \hat{V}(\hat{\mathbf{x}})(1-\hat{\eta}) & \text{in } \hat{G}_H \\ \hat{\mathbf{F}} = \hat{\eta}\hat{\mathbf{q}} & \text{in } \hat{G}_H \\ \hat{\eta} = 0 & \text{on } \hat{\Gamma}_P \cup \hat{\Gamma}_-. \end{cases} \tag{3.21}$$

As before, the total extraction rate is given by the following expression

$$\hat{J}(\hat{u}) = \int_{\hat{\Gamma}_A} \hat{\mathbf{F}} \cdot \boldsymbol{\nu} \, d\hat{\Gamma}. \tag{3.22}$$

3.2.2 Approximation of Sinks by Dirac Functions

In the case when the well screen of extraction probes is negligible with respect to the horizontal dimensions of the remediation site, the active wells can be considered as point sinks. In spite of this, the passive wells are considered as a Dirichlet boundary condition, i.e., they have positive radii. The reason is, that it is not mathematically correct to fix the solution at a point (when a passive well is described as a point) with the prescribed value. Hence, the BVP (3.1) can be approximated by the following boundary value problem with Dirac-type sinks: One has to find a function y and numbers $v_j \in \mathbb{R}$ such that

$$\begin{cases} \nabla \cdot \mathbf{q} = LT(y_R - y) - \sum_{j=1}^{N} v_j \delta_j & \text{in } G \\ \mathbf{q} = -T\nabla y & \text{in } G \\ \mathbf{q} \cdot \boldsymbol{\nu} = -lT(y_R - y) + q_N & \text{on } \Gamma_N \\ y = y_D & \text{on } \Gamma_D, \end{cases} \quad (3.23)$$

where $\delta_j = \delta(\mathbf{x} - \mathbf{x}_j)$. In the differential equation, the coefficients v_j for passive wells are unknown. They have to be determined such that they are compatible with the boundary conditions in BVP (3.1). For passive wells we require

$$\frac{1}{2\pi R_j} \int_{\Gamma_j} y \, d\Gamma = y_j \text{ for each } \Gamma_j \subset \Gamma_P.$$

For active wells we have to distinguish between the different types of conditions that are imposed.

(P) Analogously to the passive wells we require

$$\frac{1}{2\pi R_j} \int_{\Gamma_j} y \, d\Gamma = y_j \text{ for each } \Gamma_j \subset \Gamma_A$$

(F), (D) We identify

$$v_j = u_j \text{ for each } \Gamma_j \subset \Gamma_A.$$

For $|\mathbf{x} - \mathbf{x}_j|$ small, the solution y becomes negative in G_A. In this case, p and ρ are not determined. Since we need the gas density ρ in BVP (3.6), this problem cannot directly be approximated by a problem with singular wells. Nevertheless, if we restrict ourselves to small pressure differences (i.e., $\frac{p - p_R}{p_R}$ small with a reference pressure p_R), we can approximate C in the diffusive term by $\rho_R \frac{C}{\rho}$ and get a problem for the unknown $\eta = \frac{C}{\rho}$ which can be solved independently of ρ. In a similar way as before, we can approximate BVP (3.6) by

$$\begin{cases} -\nabla \cdot (D\rho_R \nabla \eta) + \rho_R \mathbf{q} \cdot \nabla \eta = V(\eta_S - \eta) - \sum_{j=1}^{n} \eta v_j \delta_j & \text{in } G \setminus \overline{G_P} \\ \mathbf{F} \cdot \boldsymbol{\nu} = \mathbf{f} \cdot \boldsymbol{\nu} & \text{on } \Gamma_+ \\ \eta = 0 & \text{on } \Gamma_- \\ \int_{\Gamma_j} \mathbf{F} \cdot \boldsymbol{\nu} \, d\Gamma = 0 & \text{for each } \Gamma_j \subset \Gamma_P. \end{cases} \quad (3.24)$$

3.2.3 Justification of a Leakage Term

Removing volatile compounds from the soil, an often encountered situation is an approximately uniformly distributed pollutant in an also uniformly conducting soil layer and a water table of constant depth d. In many cases, the soil surface is covered by a thin layer of relatively impermeable material. This means that our domain is $G_S = \{\boldsymbol{X} \in \mathbb{R}^3 : -d_N \leq z \leq d_D\}$ with Dirichlet boundary $\Gamma_D = \{\boldsymbol{X} \in \mathbb{R}^3 : z = d_D\}$ (upper boundary) and Neumann boundary $\Gamma_N = \{\boldsymbol{X} \in \mathbb{R}^3 : z = -d_N\}$ (lower boundary). The conductivity function is taken as

$$K(z) = \begin{cases} K_{min} & \text{for} \quad 0 < z \leq d_D \\ K & \text{for} \quad -d_N \leq z \leq 0 \end{cases}.$$

We realize that for a precise description of the pollutant transport we have to solve the 3-D equivalent of problem (3.23) together with equations describing the 3-D airflow $\boldsymbol{Q}$:

$$\begin{cases} \boldsymbol{Q} = -K\nabla Y & \text{in } G_S \\ \nabla \cdot \boldsymbol{Q} = -\sum_i u_i \delta(\boldsymbol{X} - \boldsymbol{X}_i) & \text{in } G_S \end{cases} \quad (3.25)$$

and appropriate boundary conditions. If we want to substitute these two 3-D problems by 2-D problems, we have to make three assumptions:

1. In the low conducting layer which covers the soil the air flow is essentially vertical, i.e., we have $Q_r = 0$ and $Q_z = -K_{min}\partial_z Y(r, z)$.

2. If the water table is high and $K_{min} \ll K$, the air flow in the soil is essentially horizontal outside a small neighborhood of the wells. For vertically constant pollutant distribution, we can vertically integrate the pollutant transport equation and get a 2-D problem where the 2-D airflow $\mathbf{q}$ is approximately the vertically integrated radial component of the 3-D airflow:

$$\mathbf{q} \approx d_N \bar{\boldsymbol{Q}}_r = \int_{-d_N}^{0} \boldsymbol{Q}_r dz = -\int_{-d_N}^{0} \nabla_r Y dz = -K d_N \nabla_r \bar{Y} = -T \nabla_r \bar{Y} \quad (3.26)$$

with the mean value $\bar{Y}(r) = \frac{1}{d_N} \int_{-d_N}^{0} Y(r, z) dz$ and the transmissivity $T = d_N K$.

3. The averaged radial air flow can approximately be computed by the 2-D problem (3.23) with a suitable leakage L.

In the following two sections we will show how the last assumption can be justified.

3-D Air Flow Problem for a Single Well

We now derive an analytic expression for the gas-flow in the case of a single point sink without insulating cover. For this, let $\boldsymbol{X}_0 = (0, 0, -z_0)$ be the location of the extraction well. Then our air flow problem in 3-D space is

$$\begin{cases} \mathbf{Q} = -K\nabla Y & \text{in } G_S \\ \nabla \cdot \mathbf{Q} = -u_0\delta(\mathbf{X} - \mathbf{X}_0) & \text{in } G_S \\ Y = Y_R & \text{on } \Gamma_D \\ \mathbf{Q} \cdot \boldsymbol{\nu} = 0 & \text{on } \Gamma_N \; . \end{cases} \qquad (3.27)$$

In the absence of an insulting cover, the solution of this problem can be easily constructed by subsequent reflections of the sink at the Dirichlet- and Neumann-boundary. We denote the locations of the mirror sinks/sources by $\mathbf{X}'_0 = -\mathbf{X}_0$, $\mathbf{X}_i = (0, 0, -z_i)$ and $\mathbf{X}'_i = (0, 0, -z'_i)$ with

$$z_i = 2di - z_0 \;, \quad z'_i = 2di + z_0 \quad \text{for } i \in \mathbb{N} \; . \qquad (3.28)$$

We get the solution by superposition of fundamental solutions for the full 3-D space

$$Y(\mathbf{X}) = Y_R - \frac{u_0}{4\pi K}\left(\frac{1}{|\mathbf{X} - \mathbf{X}_0|} - \frac{1}{|\mathbf{X} + \mathbf{X}_0|}\right)$$
$$-\frac{u_0}{4\pi K}\sum_{i=1}^{\infty}(-1)^{i+1}\left(\frac{1}{|\mathbf{X} - \mathbf{X}_i|} - \frac{1}{|\mathbf{X} + \mathbf{X}_i|} - \frac{1}{|\mathbf{X} - \mathbf{X}'_i|} + \frac{1}{|\mathbf{X} + \mathbf{X}'_i|}\right) \qquad (3.29)$$

$$\mathbf{Q}(\mathbf{X}) = -\frac{u_0}{4\pi}\left(\frac{\mathbf{X} - \mathbf{X}_0}{|\mathbf{X} - \mathbf{X}_0|^3} - \frac{\mathbf{X} + \mathbf{X}_0}{|\mathbf{X} + \mathbf{X}_0|^3}\right)$$
$$-\frac{u_0}{4\pi}\sum_{i=1}^{\infty}(-1)^{i}\left(\frac{\mathbf{X} - \mathbf{X}_i}{|\mathbf{X} - \mathbf{X}_i|^3} - \frac{\mathbf{X} + \mathbf{X}_i}{|\mathbf{X} + \mathbf{X}_i|^3} - \frac{\mathbf{X} - \mathbf{X}'_i}{|\mathbf{X} - \mathbf{X}'_i|^3} + \frac{\mathbf{X} + \mathbf{X}'_i}{|\mathbf{X} + \mathbf{X}'_i|^3}\right) \qquad (3.30)$$

Averaging the 3-D Air Flow Problem

We now want to average the full 3-D problem vertically. Introducing the mean value $\bar{Q}_r = \frac{1}{d}\int_{-d}^{0} Q_r dz$ we get

$$\int_{-d}^{0} \nabla \cdot \mathbf{Q} \, dz = \int_{-d}^{0} \frac{1}{r}\partial_r(rQ_r) dz + \int_{-d}^{0} \partial_z Q_z \, dz$$
$$= d\frac{1}{r}\partial_r(r\bar{Q}_r) + Q_z(r, z = 0) - Q_z(r, z = -d)$$
$$= d\nabla_r \cdot \bar{\mathbf{Q}}_r + Q_z(r, 0) = d\nabla_r \cdot \bar{\mathbf{Q}}_r - K\partial_z Y(r, 0) \; .$$

The derivative $\partial_z Y(r,0)$ has to be understood as the limit from below $\partial_z Y(r,0) = \lim_{z \nearrow 0} \partial_z Y(r,z)$, the same holds for $\partial_{zz}Y(r,0)$ later in the text. Using the second equation in problem (3.27) we obtain

$$d\nabla_r \cdot \bar{\mathbf{Q}}_r = K\partial_z Y(r,0) - u_0\delta(\mathbf{x}) \; .$$

3.2. Mathematical Analysis, Simplifications ...

From the interface condition for the flow at $z = 0$ we get $Y(0) = Y_R - \delta \frac{K}{K_{min}} \partial_z Y(0)$. For arbitrary r we will now expand $Y(z)$ in a power series up to terms of order z^2

$$Y(z) \approx Y_R - \delta \frac{K}{K_{min}} \partial_z Y(0) + z \partial_z Y(0) + \frac{z^2}{2} \partial_{zz} Y(0) \tag{3.31}$$

$$\partial_z Y(z) \approx \partial_z Y(0) + z \partial_{zz} Y(0) \tag{3.32}$$

From the Neumann boundary condition in problem (3.27) we get $\partial_z Y(r, -d) = 0$ and therefore $\partial_{zz} Y(0) = \frac{1}{d} \partial_z Y(0)$, which gives

$$Y(z) \approx Y_R + \left(-\delta \frac{K}{K_{min}} + z + \frac{z^2}{2d} \right) \partial_z Y(0) \;.$$

If we integrate the last equation in z we get

$$\bar{Y}(r) = \frac{1}{d} \int_{-d}^{0} Y(r, z) dz \approx Y_R - \left(\delta \frac{K}{K_{min}} + \frac{d}{3} \right) \partial_z Y(r, 0) \;. \tag{3.33}$$

This can be used to eliminate $\partial_z Y(r, 0)$ from (3.2.3), which yields the main result

$$\nabla_r \cdot \bar{Q}_r \approx \frac{1}{\delta d \frac{K}{K_{min}} + \frac{d^2}{3}} K(Y_R - \bar{Y}) - \frac{u_0}{d} \delta(\mathbf{x}) \;. \tag{3.34}$$

Finally, in (3.26) and (3.34) we perform the substitutions $d\bar{Q}_r \to \mathbf{q}$, $\bar{Y} \to y$, $Y_R \to y_R$ and identify

$$L = \left(\delta d \frac{K}{K_{min}} + \frac{d^2}{3} \right)^{-1} \;. \tag{3.35}$$

Then we get the approximated 2-D problem for the airflow

$$\begin{cases} \mathbf{q} = -T \nabla y & \text{in } \mathbb{R}^2 \\ \nabla \cdot \mathbf{q} = LT(y_R - y) - u_0 \delta(\mathbf{x}) & \text{in } \mathbb{R}^2 \\ \lim_{|\mathbf{x}| \to \infty} y = y_R \;. \end{cases} \tag{3.36}$$

Remark 3.7 We emphasize that these equations are based on the expansion (3.31) together with the Neumann boundary condition at the water table $z = -d$, which lead to the approximation (3.33) for the mean value $\bar{Y}$. Obviously, close to the well the approximation (3.33) is not valid, but for small differences $Y(r, 0) - Y(r, -d)$ one can expect that it will be true. In Figure 3.2, for a soil without cover we compare the profiles $rY(r, z)$ (solid lines) and its approximations (3.31) (dashed lines) along vertical cross-sections for several values of r. Here we haven taken the parameter values $d = 5$, $z_0 = 2.5$, $K = 1$ and $u_0 = 1$ in order to compute $Y(r, z)$ from (3.29). One can expect that for a covered soil, approximation (3.31) is even better. □

Remark 3.8 If the soil is covered by an insulating layer, formula (3.35) is of only limited value, since usually K_{min} is unknown. In the next section we will propose a method how to compute an estimated value for L from the so-called *effective radius* of a well. □

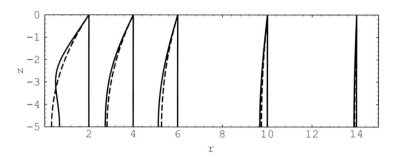

Figure 3.2: Comparison of $rY(r,z)$ (solid line) and its quadratic approximation (dashed line) for a well located at $r = 0, z = -2.5$, the water table is at $z = -5$. The dependence on z is shown for several values of r. We have $Y(2,-5) = -0.0265324$.

Solving the 2-D Airflow Equation

In this section we will present solutions of the 2-D equation (3.36) and a method to estimate the value of L. Because the problem is radially symmetric, we use polar coordinates and get

$$T \frac{1}{r} \frac{d}{dr}(r \frac{d}{dr} y) = -LT(y_R - y) \ . \tag{3.37}$$

Substituting $y(r) = w(r) + y_R$ and multiplication by $-\frac{r^2}{T}$ yields

$$r(\frac{dw}{dr} + r \frac{d^2w}{dr^2}) = Lr^2 w \ . \tag{3.38}$$

With the ansatz $w(r) = v(z)$ and $z = \sqrt{L}r$ we get a standard form of the modified Bessel differential equation of order zero (see [AS64] Formula 9.6.1):

$$z \frac{dv}{dz} + z^2 \left(\frac{d^2v}{dz^2} - v \right) = 0 \tag{3.39}$$

with the general solution

$$v(z) = C_1 I_0(z) + C_2 K_0(z) \ , \tag{3.40}$$

where I_0, K_0 denote the modified Bessel functions, see Figure 3.3. From the condition $\lim_{z \to \infty} v(z) = 0$ we get $C_1 = 0$. Backward substitution yields

$$y(r) = y_R + C_2 K_0(\sqrt{L}r) \ . \tag{3.41}$$

For the radial component q_r of the gas-flow we get

$$q_r(r) = -T \frac{d}{dr} y(r) = C_2 T \sqrt{L} K_1(\sqrt{L}r) \ . \tag{3.42}$$

Because of $\lim_{r \to 0} r K_1(ar) = \frac{1}{a}$ and $\lim_{r \to 0} 2\pi r q_r(r) = -u_0$ we get $C_2 = -\frac{u_0}{2\pi T}$ and

$$y(r) = y_R - \frac{u_0}{2\pi T} K_0(r\sqrt{L}) \quad \text{and} \quad q_r(r) = -\frac{u_0}{2\pi} \sqrt{L} K_1(r\sqrt{L}) \ . \tag{3.43}$$

3.2. Mathematical Analysis, Simplifications ...

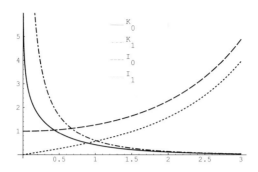

Figure 3.3: Modified Bessel functions K_0, K_1, I_0 and I_1..

The last equation show the fundamental solution of the radial symmetric air flow problem with a leakage term for a single well (see also Remark 4.7).

As already derived, we have $L = \left(\delta d \frac{K}{K_{\min}} + \frac{d^2}{3}\right)^{-1}$. Since in practical cases $K_{\min}$ cannot be determined (cracks in the asphalt layer etc.), we need a different approach in order to estimate L. This can be done by means of the effective radius R_{eff} of a well, i.e., the radius of a circle through which 90% of the air enters the soil. Usually, this quantity can be derived from field measurements.

The surface seal has an influence on the effective radius. The presence of an impermeable cover increases the range of influence (cf. Hiller [Hil]). Hiller discussed the dependence of the range of influence for a single well with a prescribed under-pressure on the well position, on the well screen and on the depth of the vadose zone. The effective radius depends on the depth of the water table and the depth of the pumping well. Formulas for infinite depth and a homogeneous soil matrix can be found in Wilson [Wil95]. They reveal a strong decrease of the effective radius for a high water table. For the analytical description of single-phase flow caused by a single extraction well for a perfectly layered subsurface we refer the reader to Nieuwenhuizen, Zijl and Van Veldhuizen [NZV95]. Schumacher and Slodička [SS97] studied the influence of the heterogeneity, well position and the depth of the water table on the effective radius and on the pressure at the extraction well.

For a well without cover, it can also be determined from the equation

$$\int_{|\mathbf{x}| \leq R_{\text{eff}}} Q_z(\mathbf{x}, 0) d\mathbf{x} = -0.9 u_0$$

where $\mathbf{Q}(\mathbf{x},0)$ is given by (3.30). For our 2-D simplification, this radius depends only on L; one has

$$R_{\text{eff}} = \frac{z_{10}}{\sqrt{L}} \qquad (3.44)$$

where $z_{10} = 3.21432$ solves $z_{10} K_1(z_{10}) = \frac{1}{10}$. We choose L such that both the 3-D- and the 2-D-solutions have the same effective radius:

$$2\pi R_{\text{eff}} \bar{Q}_r(R_{\text{eff}}) = 2\pi R_{\text{eff}} \, q(R_{\text{eff}}) = -0.1 u_0 \ . \qquad (3.45)$$

A comparison of the 2-D and 3-D radial flow is shown in Figure 3.4. The parameter values are $u_0 = 1$, $d = 5$, $z_0 = 2.5$ resulting in an effective radius $R_{\text{eff}} \approx 9.85158$ and $L \approx 0.106455$, compared to the theoretical value $L = \frac{3}{d^2} = 0.12$. Additionally, the figure shows the radial 3-D flow for a vertical line well uniformly distributed between $-4 \leq z_0 \leq -1$.

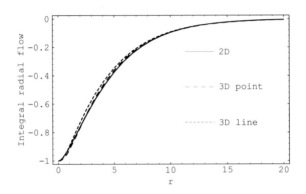

Figure 3.4: Integral radial flow for a 2-D and 3-D point well and a 3-D line well.

3.2.4 Streamline Method

The pollutant flow problem by soil venting is convection dominated, i.e., the dissemination of the VOCs by diffusion is negligible with respect to the convection created by air pumping wells. The streamline method is a simple but effective method how to compute the NAPL extraction rate for transport problems. Now we explain very briefly the main idea of this approach (see also Section 7.2).

If we assume $D = 0$, which seems to be close to reality for stationary conditions, we can use the methods of streamlines for the calculation of concentrations along the boundary of the active holes. This means, that we first have to compute the streamlines of the flux and then solve a 1-D initial value problem along each streamline. In more detail, the streamlines are given by the initial value problem

$$\frac{d}{dt}\mathbf{x}(t) = \mathbf{q}, \quad \mathbf{x}(t_A) = \mathbf{x}_A \qquad (3.46)$$

We assume that $\mathbf{x}_A \in \Gamma_A$, which means that the end point is on the boundary of an active well. Solving this ODE backwards in time, we find a parameter value $t = t_1 < t_A$ for which the streamline crosses the boundary (of a passive well or the exterior boundary) for the first time, called the *boundary case*, or where the modulus of the flow vector $\mathbf{q}$ is less than a prescribed threshold, called the *interior case*.

After determining a streamline, we consider the contaminant transport along that streamline. Using equations (3.6) and (3.1) we get for the relative mass concentration

3.2. Mathematical Analysis, Simplifications ...

η along the streamline

$$\frac{d}{dt}\eta = \nabla \eta \cdot \mathbf{q} = V(\eta_S - \eta) - \eta \nabla \cdot \mathbf{q}$$
$$= V(\eta_S - \eta) - \eta LT(y_R - y) \quad (3.47)$$

together with

$$\eta(t_0) = \begin{cases} 0, & \text{for the boundary case} \\ \eta_{\text{eq}}, & \text{for the interior case} \end{cases}$$

as an initial value problem. Here, η_{eq} is the equilibrium saturation given as the solution of the equation

$$V(\eta_S - \eta_{\text{eq}}) - \eta_{\text{eq}} LT(y_R - y) = 0.$$

The solution $\eta(t_A)$ gives the desired concentration at the point $(x_1(t_A), x_2(t_A))$ on the boundary of the active well. The total pollutant extraction rate can be determined by numerical integration of $\mathbf{f} \cdot \boldsymbol{\nu}$ along Γ_A, i.e., by evaluating

$$\int_{\Gamma_A} \mathbf{f} \cdot \boldsymbol{\nu} \, d\Gamma = \sum_j \int_{\Gamma_j} \mathbf{f} \cdot \boldsymbol{\nu} \, d\Gamma.$$

The NAPL concentration changes very rapidly near the extraction well, thus a clever integration formula must be used by evaluation of the extraction rate by probes (see Section 7.3).

Let us note that the *streamline method* in dimensionless form (see Section 3.2.1) consists of solving the system

$$\boxed{\tfrac{d}{dt}\hat{\mathbf{x}} = \hat{\mathbf{q}}, \quad \tfrac{d}{dt}\hat{\eta} = \hat{V}(1 - \hat{\eta}) + \hat{L}\hat{T}\hat{y}\hat{\eta}}. \quad (3.48)$$

We will use this formulation later by some examples.

Chapter 4

Well-Posedness and Optimality

Mathematical modeling of soil venting leads to a system of partial differential equations. Before starting all computations, one must do the qualitative and quantitative analysis of solutions for corresponding problems (at least in simple situations as the study cases).

The main goal of this chapter is to describe the mathematical foundation for the existence, uniqueness and some regularity properties for solution of the air and pollutant flow problem. The cases of smooth and non smooth transmissivity are discussed. In both situations the definition of a solution is given (see Definitions 4.5 and 4.9). Theorem 4.11 gives the comparison of both definitions.

4.1 Air Flow in a Domain with Holes

Problem 4.1 (Air flow) *We choose the discharge boundary condition [D]. Let $u = (u_1, \ldots, u_n)$ be given. Find $y \to \mathbb{R}$ and $\gamma = (\gamma_1, \ldots, \gamma_n)$ such that*

$$\begin{cases} \nabla \cdot \mathbf{q} = LT(y_R - y) & \text{in } G_H \\ \mathbf{q} = -T\nabla y & \text{in } G_H \\ \mathbf{q} \cdot \boldsymbol{\nu} = -lT(y_R - y) + q_N & \text{on } \Gamma_N \\ y = y_D & \text{on } \Gamma_D \cup \Gamma_P \\ y = \gamma_j & \text{on each } \Gamma_j \subset \Gamma_A \\ \int_{\Gamma_j} \mathbf{q} \cdot \boldsymbol{\nu} \, d\Gamma = u_j & \text{for each } \Gamma_j \subset \Gamma_A. \end{cases} \quad (4.1)$$

Theorem 4.2 (Existence and uniqueness) *Let G_H be a bounded open domain in $\mathbb{R}^2$ with a piecewise C^1 boundary Γ such that $\int_{\Gamma_D \cup \Gamma_P} d\Gamma > 0$. Let y_D be continuous, $y_R, q_N \in L^2(\Gamma_N)$. We suppose that L, l, T are continuous and $0 < C_0 < T(\mathbf{x}) < C_1$, $0 \le L(\mathbf{x}), l(\mathbf{x}) < C_1$ where C_0 and C_1 are two constants. Then Problem 4.1 has a unique solution.*

Proof: The mapping from $\mathbb{R}^n$ to $\mathbb{R}^n$ defined by

$$u_j = \int_{\Gamma_j} \mathbf{q}(\gamma) \cdot \boldsymbol{\nu} \, d\Gamma. \quad (4.2)$$

is bijective: For any set of scalars $u_j, j = 1, \ldots, n$, there exists one unique set $\gamma_j, j = 1, \ldots, n$ such that $u_j = \int_{\Gamma_j} \mathbf{q}(\gamma) \cdot \boldsymbol{\nu}$.

We see this in the following way. The mapping (4.2) is well defined, i.e., for any given set of scalars $\gamma_j, j = 1, \ldots, n$, the BVP (4.1) has one unique solution y in $W^{1,2}(G_H)$ and the normal component of the flux $\mathbf{q}$ on Γ_j ($j = 1, \ldots, n$) is in $L^2(\Gamma_j)$. Since $\mathbf{q}$ is affine with respect to γ, the mapping (4.2) is also affine from $\mathbb{R}^n$ to $\mathbb{R}^n$. Therefore, it remains to be shown that it is injective. Let γ and γ' be two vectors from $\mathbb{R}^n$, y and y' their respective solutions of the BVP (4.1) and $\mathbf{q}$ and $\mathbf{q}'$ the respective fluxes, such that

$$\int_{\Gamma_j} \mathbf{q} \cdot \boldsymbol{\nu} \, d\Gamma = \int_{\Gamma_j} \mathbf{q}' \cdot \boldsymbol{\nu} \, d\Gamma$$

for $j = 1, \ldots, n$. The solutions y and y' are in $W^{1,2}(G_H)$ then using a duality argument we can apply the following Green formula

$$0 = \int_{G_H} -(y-y')(\nabla \cdot T\nabla(y-y')) \, dx + \int_{G_H} LT(y-y')^2 \, dx$$
$$= \int_{G_H} T\nabla(y-y') \cdot \nabla(y-y') \, dx - \int_{\Gamma_A} (y-y')T\nabla(y-y') \cdot \boldsymbol{\nu} \, d\Gamma$$
$$+ \int_{G_H} LT(y-y')^2 \, dx + \int_{\Gamma_N} lT(y-y')^2 \, d\Gamma.$$

Since $y - y'$ is constant on Γ_j, we pull it outside the integral in the second term and get $\nabla(y - y') = 0$ in $L^2(G_H)$ and $y - y' \in H^1(G_H)$. From this we see that $y - y'$ is constant on the connected domain G_H. Since $y - y'$ is constant and $y - y' = 0$ on $\Gamma_D \cup \Gamma_P$, we obtain $y - y' = 0$ in G_H. □

Remark 4.3 The last theorem carries over to the situation in which the wells operate in groups (see the Remark at the end of Section 3.1.1). □

4.2 Air Flow with Dirac-Type Sinks

Problem 4.4 (Air flow) Let $u = (u_1, \ldots, u_n)$ be given. Find $y : G \to \mathbb{R}$ such that

$$\begin{cases} \nabla \cdot \mathbf{q} = LT(y_R - y) - \sum_{j=1}^n u_j \delta_j & \text{in } G \setminus \overline{G_P} \\ \mathbf{q} = -T\nabla y & \text{in } G \setminus \overline{G_P} \\ \mathbf{q} \cdot \boldsymbol{\nu} = -lT(y_R - y) + q_N & \text{on } \Gamma_N \\ y = y_D & \text{on } \Gamma_D \cup \Gamma_P. \end{cases}$$

where $\delta_j(\mathbf{x}) = \delta(\mathbf{x} - \mathbf{x}_j)$, $j = 1, \ldots, n$.

4.2. Air Flow with Dirac-Type Sinks

Let us denote the transmissivities at the active wells by T_j ($j = 1, \ldots, n$), i.e., $T_j = T(\mathbf{x}_j)$. The right-hand side of the air flow problem contains Dirac functions. In order to subtract singularities we replace δ_j ($j = 1, \ldots, n$) using

$$-\nabla \cdot (T_j \nabla Y_j) = -u_j \delta_j,$$

where

$$Y_j(\mathbf{x}) = \frac{u_j}{2\pi T_j} \ln |\mathbf{x} - \mathbf{x}_j|. \qquad (4.3)$$

Definition 4.5 (Weak solution) *We say that $y \in W^{1,p}(G \setminus \overline{G_P})$, $1 < p < 2$ is a "weak" solution of Problem 4.4 iff*

1. $y = \tilde{y} + \sum_{j=1}^{n} Y_j$ *and*

2. $\tilde{y}$ *is the solution of the problem*

$$\begin{cases}
-\nabla \cdot (T \nabla \tilde{y}) + LT\tilde{y} = LT\left(y_R - \sum_{j=1}^{n} Y_j\right) \\
\qquad\qquad + \sum_{j=1}^{n} \nabla \cdot ((T - T_j) \nabla Y_j) & \text{in } G \setminus \overline{G_P} \\
-T\nabla \tilde{y} \cdot \boldsymbol{\nu} - lT\tilde{y} = -lT\left(y_R - \sum_{j=1}^{n} Y_j\right) + q_N & \\
\qquad\qquad + \sum_{j=1}^{n} T\nabla Y_j \cdot \boldsymbol{\nu} & \text{on } \Gamma_N \\
\tilde{y} = y_D - \sum_{j=1}^{n} Y_j & \text{on } \Gamma_D \cup \Gamma_P.
\end{cases} \qquad (4.4)$$

Theorem 4.6 (Existence and uniqueness) *Let G be a bounded open domain in $\mathbb{R}^2$ with a Lipschitz continuous boundary Γ. Let y_D be continuous, $q_N \in L^2(\Gamma_N)$, $0 < C_0 < T(\mathbf{x}) < C_1$, $0 \leq L(\mathbf{x}), l(\mathbf{x}) < C_1$ and T is supposed to be Hölder continuous at $\mathbf{x}_j$ near each active well ($j = 1, \ldots, n$). Then there exists a unique solution of Problem 4.4.*

Proof: Using (4.3) we can write

$$\nabla Y_j = \frac{u_j}{2\pi T_j} \frac{\mathbf{x} - \mathbf{x}_j}{|\mathbf{x} - \mathbf{x}_j|^2}.$$

The Hölder continuity of the transmissivity at $\mathbf{x}_j$ near each active well implies

$$(T - T_j) \nabla Y_j \in L^p(G \setminus \overline{G_P}) \times L^p(G \setminus \overline{G_P})$$

($j = 1, \ldots, n$) for some $p > 2$. The existence of $\tilde{y} \in W^{1,2}(G \setminus \overline{G_P})$ follows from the theory of linear elliptic equations (cf. Gilbarg-Trudinger [GT83]), and $y = \tilde{y} + \sum_{j=1}^{n} Y_j$.

According to Definition 4.5 the uniqueness of y is equivalent to the uniqueness of $\tilde{y}$ which is guaranteed by the ellipticity of the operator $\mathcal{A}(w) = -\nabla(T\nabla w) + LTw$. $\square$

Remark 4.7 If $L > 0$, the active wells can be modeled in another way using modified Bessel function K_0 (see [AS64] Section 9.6). Let us denote by $L_j (j = 1, \ldots, n)$ the values of the leakage term at the active wells, i.e., $L_j = L(\mathbf{x}_j)$ and define Y_j $(j = 1, \ldots, n)$ as

$$Y_j(\mathbf{x}) = y_R - \frac{u_j}{2\pi T_j} K_0 \left(|\mathbf{x} - \mathbf{x}_j| \sqrt{L_j} \right). \tag{4.5}$$

Then Y_j $(j = 1, \ldots, n)$ are the fundamental solutions of

$$-\nabla \cdot (T_j \nabla Y_j) - L_j T_j (y_R - Y_j) = -u_j \delta_j.$$

For their gradients we have

$$\nabla Y_j = \frac{u_j}{2\pi T_j} \frac{\mathbf{x} - \mathbf{x}_j}{|\mathbf{x} - \mathbf{x}_j|} \sqrt{L_j} K_1 \left(|\mathbf{x} - \mathbf{x}_j| \sqrt{L_j} \right),$$

with the modified Bessel function K_1. Analogously to (4.4) we can write

$$\begin{cases} -\nabla \cdot (T \nabla \tilde{y}) + LT\tilde{y} = LT \left(y_R - \sum_{j=1}^n Y_j \right) + \sum_{j=1}^n \nabla \cdot ((T - T_j) \nabla Y_j) \\ \qquad\qquad\qquad - \sum_{j=1}^n L_j T_j (y_R - Y_j) & \text{in } G \setminus \overline{G_P} \\ -T \nabla \tilde{y} \cdot \boldsymbol{\nu} - lT\tilde{y} = -lT \left(y_R - \sum_{j=1}^n Y_j \right) + q_N + \sum_{j=1}^n T \nabla Y_j \cdot \boldsymbol{\nu} & \text{on } \Gamma_N \\ \tilde{y} = y_D - \sum_{j=1}^n Y_j & \text{on } \Gamma_D \cup \Gamma_P. \end{cases} \tag{4.6}$$

According to [AS64] Section 9.6 we have

$$K_0(z) \sim -\ln z, \qquad K_1(z) \sim \frac{1}{z}, \qquad \text{as } z \to 0,$$

thus the existence of the solution $y = \tilde{y} + \sum_{j=1}^n Y_j$ can be proved in the same way as for Y_j $(j = 1, \ldots, n)$ defined by (4.3). □

Theorem 4.8 (Regularity) *We suppose $l = 0$. Let $y_R, T, L \in C^\infty(\overline{G_H})$ be such that $0 < C_0 < T(\mathbf{x}) < C_1$, $0 \le L(\mathbf{x}) < C_1$ Then*

(i) $y \in C^\infty(G_H)$.

(ii) *Let the boundary of the G_H be of class C^∞. Let us consider Dirichlet (Neumann) boundary conditions. If y_D (q_N) $\in C^\infty$ on the whole boundary ∂G_H, then $y \in C^\infty(\overline{G_H})$.*

Proof: Let us note that $y = \tilde{y} + \sum_{j=1}^n Y_j$. Each function $Y_j (j = 1, \ldots, n)$ belongs to $C^\infty(\overline{G_H})$. According to the fact that $y_R, T, L \in C^\infty(\overline{G_H})$ one can see that

$$LT \left(y_R - \sum_{j=1}^n Y_j \right) + \sum_{j=1}^n \nabla \cdot ((T - T_j) \nabla Y_j)$$

belongs to $C^\infty(\overline{G_H})$. Now we can apply Dautray, Lions [DL90] (p. 585, Proposition 5) in order to complete the proof. □

4.3 Non-smooth Transmissivity at Active Wells

Throughout this section we suppose that G is a bounded open domain in $\mathbb{R}^2$ with a Lipschitz continuous boundary Γ, y_D is continuous, $q_N \in L^2(\Gamma_N)$, $0 < C_0 < T(\mathbf{x}) < C_1$, $0 \leq L(\mathbf{x})$ and $l(\mathbf{x}) < C_1$.

The crucial assumption of the last paragraph was the local Hölder continuity of the transmissivity at the active wells. Then we were able to give a reasonable definition of the solution. If the transmissivity has jumps at the active wells, we must give a definition of solution different from Definition 4.5.

We want to solve Problem 4.4. Using the method of superposition ($y = y^1 + y^2$) we first solve

$$\begin{cases} -\nabla \cdot (T\nabla y^1) + LTy^1 = LTy_R & \text{in } G \setminus \overline{G_P}, \\ -T\nabla y^1 \cdot \boldsymbol{\nu} - lTy^1 = -lTy_R + q_N & \text{on } \Gamma_N, \\ y^1 = y_D & \text{on } \Gamma_D \cup \Gamma_P. \end{cases} \quad (4.7)$$

This problem admits a unique solution $y^1 \in W^{1,2}(G \setminus \overline{G_P})$. Now it is sufficient to find the solution of the following problem

$$\begin{cases} -\nabla \cdot (T\nabla y^2) + LTy^2 = -\sum_{j=1}^n u_j \delta_j & \text{in } G \setminus \overline{G_P}, \\ -T\nabla y^2 \cdot \boldsymbol{\nu} - lTy^2 = 0 & \text{on } \Gamma_N, \\ y^2 = 0 & \text{on } \Gamma_D \cup \Gamma_P. \end{cases} \quad (4.8)$$

First, we consider the adjoint problem with the right-hand side $F \in L^2(G \setminus \overline{G_P})$

$$\begin{cases} -\nabla \cdot (T\nabla Y_F^2) + LTY_F^2 = F & \text{in } G \setminus \overline{G_P} \\ -T\nabla Y_F^2 \cdot \boldsymbol{\nu} - lTY_F^2 = 0 & \text{on } \Gamma_N \\ Y_F^2 = 0 & \text{on } \Gamma_D \cup \Gamma_P. \end{cases} \quad (4.9)$$

This has a unique solution $Y_F^2 \in W^{1,2}(G \setminus \overline{G_P})$. Thus we can define the linear operator $A \in \mathcal{L}(L^2(G \setminus \overline{G_P}), W^{1,2}(G \setminus \overline{G_P}))$ ($\mathcal{L}(X, Y)$ denotes the space of all linear mappings from X to Y) as

$$A: \quad F \longrightarrow Y_F^2.$$

Moreover if $Y_F^2 \in C^0(\overline{G \setminus \overline{G_P}})$, i.e., $A \in \mathcal{L}(L^2(G \setminus \overline{G_P}), C^0(\overline{G \setminus \overline{G_P}}))$, then for the dual operator A^* we have $A^* \in \mathcal{L}((C^0(\overline{G \setminus \overline{G_P}}))^*, L^2(G \setminus \overline{G_P}))$. Using the prolongation operator (see Kufner, John, Fučik [KJF77] Theorem 1.7.5) we can write $A^* \in \mathcal{L}(\mathcal{M}_b, L^2(G \setminus \overline{G_P}))$, where $\mathcal{M}_b$ is the space of Borel measures. Thus, applying the transposition method of Stampacchia [Sta66] we can define the function $y^2 \in L^2(G \setminus \overline{G_P})$ by

$$\int_{G \setminus \overline{G_P}} y^2 F = -\int_{G \setminus \overline{G_P}} y^2 \nabla \cdot (T\nabla Y_F^2) + \int_{G \setminus \overline{G_P}} LTy^2 Y_F^2$$

$$= -\int_{G \setminus \overline{G_P}} \nabla \cdot (T\nabla y^2) Y_F^2 + \int_{G \setminus \overline{G_P}} LTy^2 Y_F^2$$

$$= -\int_{G \setminus \overline{G_P}} \sum_{j=1}^n u_j \delta_j Y_F^2 = -\sum_{j=1}^n u_j Y_F^2(\mathbf{x}_j)$$

for an arbitrary $F \in L^2(G \setminus \overline{G_P})$. At the end, using the principle of superposition, we have obtained the following definition of a "very weak" solution ("very weak" solution means that its regularity is worse than the regularity of a "weak" solution).

Definition 4.9 (Very weak solution) *We say that $y \in L^2(G \setminus \overline{G_P})$ is a "very weak" solution of Problem 4.4 iff*

1. $y = y^1 + y^2$

2. y^1 *is the solution of (4.7)*

3.
$$\int_{G \setminus \overline{G_P}} y^2 F = -\sum_{j=1}^n u_j Y_F^2(\mathbf{x}_j)$$

for all $F \in L^2(G \setminus \overline{G_P})$, where $Y_F^2 \in C^0(\overline{G \setminus \overline{G_P}})$ is the solution of the adjoint problem (4.9).

Remark 4.10

- This definition requires $Y_F^2 \in C^0(\overline{G \setminus \overline{G_P}})$ for all $F \in L^2(G \setminus \overline{G_P})$. For the Dirichlet problem this follows from Gilbarg Trudinger [GT83] Theorem 8.30.

- In this case jumps of transmissivity at the active wells are allowed.

- Existence and uniqueness of a "very weak" solution of Problem 4.4 are equivalent to existence and uniqueness of a solution of (4.8), because of $y^1 \in W^{1,2}(G \setminus \overline{G_P})$ is unique.

- The existence of $y^2 \in L^2(G \setminus \overline{G_P})$ follows from the considerations above.

- If $y^{2,1}, y^{2,2}$ are two solutions of (4.8), then

$$\int_\Omega (y^{2,1} - y^{2,2}) F = 0 \quad \forall F \in L^2(G \setminus \overline{G_P})$$

which implies uniqueness. $\square$

Theorem 4.11 (Comparison of definitions) *Let the transmissivity T be Hölder continuous near each active well. Let Y_F^2 be the solution of (4.9). We suppose that $Y_F^2 \in W^{1,p}(G \setminus \overline{G_P})$, $p > 2$, for all $F \in L^2(G \setminus \overline{G_P})$.*

Then each "weak" solution of Problem 4.4 is a "very weak" solution.

Proof: Let us note that for $p > 2$ we have $W^{1,p}(G \setminus \overline{G_P}) \hookrightarrow C^0(\overline{G \setminus \overline{G_P}})$. Problem 4.4 is linear and problem (4.7) has a unique solution $y^1 \in W^{1,2}(G \setminus \overline{G_P})$. Thus, using the method of superposition it is sufficient to compare both definitions for problem (4.8), only. Let y^2 be the solution of (4.8) in the sense of Definition 4.5, i.e.,

$$y^2 = \tilde{y^2} + \sum_{j=1}^n Y_j^2.$$

4.3. Non-smooth Transmissivity at Active Wells

Let us define the following functions $(j = 1, \ldots, n;\ \varepsilon > 0)$

$$Y_j^{2,\varepsilon}(\mathbf{x}) = \begin{cases} Y_j^2(\mathbf{x}) & |\mathbf{x} - \mathbf{x}_j| > \varepsilon \\ -\dfrac{u_j}{2\pi T_j}\ln\varepsilon & |\mathbf{x} - \mathbf{x}_j| \leq \varepsilon. \end{cases}$$

We can see that $Y_j^{2,\varepsilon} \in W^{1,2}(G \setminus \overline{G_P})$ and for $\varepsilon \to 0$ we have

$$\begin{aligned} Y_j^{2,\varepsilon} &\to Y_j^2, & \nabla Y_j^{2,\varepsilon} &\to \nabla Y_j^2 & \text{almost everywhere in } G \setminus \overline{G_P} \\ Y_j^{2,\varepsilon} &\to Y_j^2 & & & \text{in } W^{1,q}(G \setminus \overline{G_P}) \text{ for } 1 < q < 2 \end{aligned} \quad (4.10)$$

Now, we have to check the integral identity $(F \in L^2(G \setminus \overline{G_P}))$

$$\int_{G \setminus \overline{G_P}} y^2 F = -\sum_{j=1}^n u_j Y_F^2(\mathbf{x}_j) \quad (4.11)$$

from Definition 4.9. According to (4.10) we can write

$$\int_{G \setminus \overline{G_P}} y^2 F = \int_{G \setminus \overline{G_P}} \left(\tilde{y}^2 + \sum_{j=1}^n Y_j^2\right) F = \lim_{\varepsilon \to 0} \int_{G \setminus \overline{G_P}} \left(\tilde{y}^2 + \sum_{j=1}^n Y_j^{2,\varepsilon}\right) F. \quad (4.12)$$

Using the fact that $\tilde{y}^2 + \sum_{j=1}^n Y_j^{2,\varepsilon} \in W^{1,2}(G \setminus \overline{G_P})$ and the relation (4.9) we have

$$\begin{aligned} \lim_{\varepsilon \to 0} \int_{G\setminus\overline{G_P}} \left(\tilde{y}^2 + \sum_{j=1}^n Y_j^{2,\varepsilon}\right) F &= \lim_{\varepsilon \to 0} \int_{G\setminus\overline{G_P}} T\nabla\left(\tilde{y}^2 + \sum_{j=1}^n Y_j^{2,\varepsilon}\right)\nabla Y_F^2 \\ + \lim_{\varepsilon \to 0} \int_{G\setminus\overline{G_P}} LTY_F^2\left(\tilde{y}^2 + \sum_{j=1}^n Y_j^2\right) &+ \lim_{\varepsilon \to 0} \int_{\partial(G\setminus\overline{G_P})\cap\Gamma_N} lTY_F^2\left(\tilde{y}^2 + \sum_{j=1}^n Y_j^2\right) \\ \int_{G\setminus\overline{G_P}} T\nabla\tilde{y}^2\nabla Y_F^2 &+ \int_{G\setminus\overline{G_P}} LTY_F^2\tilde{y}^2 + \int_{\partial(G\setminus\overline{G_P})\cap\Gamma_N} lTY_F^2\tilde{y}^2 \\ + \sum_{j=1}^n \lim_{\varepsilon \to 0}\left[\int_{G\setminus\overline{G_P}} T\nabla Y_j^{2,\varepsilon}\nabla Y_F^2 + \int_{G\setminus\overline{G_P}} LTY_F^2 Y_j^{2,\varepsilon} + \int_{\partial(G\setminus\overline{G_P})\cap\Gamma_N} lTY_F^2 Y_j^{2,\varepsilon}\right]. \end{aligned} \quad (4.13)$$

Now, applying (4.10) and passing the limit for $\varepsilon \to 0$ we obtain

$$\begin{aligned} \lim_{\varepsilon\to 0}\int_{G\setminus\overline{G_P}} T\nabla Y_j^{2,\varepsilon}\nabla Y_F^2 &\longrightarrow \int_{G\setminus\overline{G_P}} T\nabla Y_j^2\nabla Y_F^2, \\ \lim_{\varepsilon\to 0}\int_{G\setminus\overline{G_P}} LTY_F^2 Y_j^{2,\varepsilon} &\longrightarrow \int_{G\setminus\overline{G_P}} LTY_F^2 Y_j^2, \\ \lim_{\varepsilon\to 0}\int_{\partial(G\setminus\overline{G_P})\cap\Gamma_N} lTY_F^2 Y_j^{2,\varepsilon} &= \int_{\partial(G\setminus\overline{G_P})\cap\Gamma_N} lTY_F^2 Y_j^2. \end{aligned} \quad (4.14)$$

Thus, (4.12)–(4.14) give

$$\begin{aligned} \int_{G\setminus\overline{G_P}} y^2 F &= \int_{G\setminus\overline{G_P}} T\nabla\tilde{y}^2\nabla Y_F^2 + \int_{G\setminus\overline{G_P}} LTY_F^2\tilde{y}^2 + \int_{\partial(G\setminus\overline{G_P})\cap\Gamma_N} lTY_F^2\tilde{y}^2 \\ &+ \sum_{j=1}^n \int_{G\setminus\overline{G_P}} T\nabla Y_j^2\nabla Y_F^2 + \sum_{j=1}^n \int_{G\setminus\overline{G_P}} LTY_F^2 Y_j^2 + \sum_{j=1}^n \int_{\partial(G\setminus\overline{G_P})\cap\Gamma_N} lTY_F^2 Y_j^2. \end{aligned} \quad (4.15)$$

Definition 4.5 (mainly its part concerning $\tilde{y}^2$) together with the Green theorem imply

$$\int_{G\setminus\overline{G_P}} T\nabla\tilde{y}^2 \nabla Y_F^2 + \int_{G\setminus\overline{G_P}} LTY_F^2\tilde{y}^2 + \int_{\partial(G\setminus\overline{G_P})\cap\Gamma_N} lTY_F^2\tilde{y}^2$$
$$+\sum_{j=1}^{n}\left[\int_{G\setminus\overline{G_P}} T\nabla Y_j^2\nabla Y_F^2 + \int_{G\setminus\overline{G_P}} LTY_F^2 Y_j^2 + \int_{\partial(G\setminus\overline{G_P})\cap\Gamma_N} lTY_F^2 Y_j^2\right]$$
$$=\sum_{j=1}^{n}\int_{G\setminus\overline{G_P}} T_j\nabla Y_j^2\nabla Y_F^2 - \sum_{j=1}^{n}\int_{\partial(G\setminus\overline{G_P})\cap\Gamma_N} T_j\nabla Y_j^2\cdot\nu Y_F^2$$
$$=\sum_{j=1}^{n}\int_{G\setminus\overline{G_P}} -\nabla\cdot(T_j\nabla Y_j^2) Y_F^2 = -\sum_{j=1}^{n}u_j\int_{G\setminus\overline{G_P}}\delta_j Y_F^2 = -\sum_{j=1}^{n}u_j Y_F^2(\mathbf{x}_j). \quad (4.16)$$

At the end, the relations (4.15) and (4.16) imply (4.11), i.e., y^2 is a "very weak" solution of problem (4.8). □

Remark 4.12 We have made the following assumption in this theorem

$$Y_F^2 \in W^{1,p}(G\setminus\overline{G_P}), \quad p > 2, \quad \forall F \in L^2(G\setminus\overline{G_P}).$$

This regularity result for Dirichlet problem follows from Meyers [Mey63] and Simader [Sim72] p. 90. □

Theorem 4.11 shows that if the transmissivity T is locally continuous near the active wells, then both definitions of solutions ("weak" and "very weak") are equivalent. This follows from the unicity of solution. Local continuity of transmissivity allows subtraction of singularity near the sinks and to apply the well known numerical schemes for computations. For the case when jumps of transmissivity on the wells are allowed we refer the reader according numerical scheme to Slodička [Slo97], [Slo98].

4.4 Contaminant Transport in a Domain with Holes

The BVP (3.6) can be written in variational form as follows.

Problem 4.13 (Contaminant transport) *Let $\mathbf{q}$ be given. Find a function $\eta : G_H \to \mathbb{R}$ such that*

$$\int_{G_H} D\rho_R \nabla\eta\cdot\nabla\varphi\,dx - \int_{G_H}\eta\mathbf{q}\cdot\nabla\varphi\,dx + \int_{\Gamma_A\cup\Gamma_+}\mathbf{q}\cdot\nu\eta\varphi\,d\Gamma + \int_{G_H} V\eta\varphi\,dx$$
$$= \int_{G_H} V\eta_S\varphi\,dx$$

holds for all $\varphi \in W^{1,2}(G_H)$ with $\varphi = 0$ on $\Gamma_P \cup \Gamma_-$.

Theorem 4.14 (Existence and uniqueness) *Let G be a bounded open domain in $\mathbb{R}^2$ with a Lipschitz continuous boundary Γ. Let $0 < D_0 < D(\mathbf{x}) < D_1$, $0 < V_0 < V(\mathbf{x}) < V_1$ in G_H, $\eta_S \in L^2(G_H)$ and $\nabla\cdot\mathbf{q} = LT(y_R - y) \geq 0$ a.e. in G_H, $\|\mathbf{q}\|_{L^\infty(G_H)} \leq C$. Then there exists a unique solution of Problem 4.13.*

4.4. Contaminant Transport in a Domain with Holes

Proof: For arbitrary $\varphi \in W^{1,2}(G_H)$, $\varphi = 0$ on $\Gamma_P \cup \Gamma_-$ we can write

$$-\int_{G_H} \eta \mathbf{q} \cdot \nabla\varphi \, dx = \int_{G_H} \nabla(\eta\mathbf{q}) \varphi \, dx - \int_{\Gamma_A \cup \Gamma_P \cup \Gamma_+ \cup \Gamma_-} \mathbf{q} \cdot \boldsymbol{\nu} \eta \varphi \, d\Gamma$$

$$= \int_{G_H} \nabla\eta \cdot \mathbf{q}\varphi \, dx + \int_{G_H} (\nabla \cdot \mathbf{q})\eta\varphi \, dx - \int_{\Gamma_A \cup \Gamma_+} \mathbf{q} \cdot \boldsymbol{\nu} \eta \varphi \, d\Gamma$$

$$= \int_{G_H} \nabla\eta \cdot \mathbf{q}\varphi \, dx - \int_{\Gamma_A \cup \Gamma_+} \mathbf{q} \cdot \boldsymbol{\nu} \eta \varphi \, d\Gamma + \int_{G_H} LT(y_R - y)\eta\varphi \, dx.$$

Setting $\varphi = \eta$ we obtain

$$-\int_{G_H} \eta \mathbf{q} \cdot \nabla\eta \, dx = -\frac{1}{2}\int_{\Gamma_A \cup \Gamma_+} \mathbf{q} \cdot \boldsymbol{\nu} \eta^2 \, dx + \frac{1}{2}\int_{G_H} LT(y_R - y)\eta^2 \, dx.$$

Using this we can write

$$\int_{G_H} D\rho_R \nabla\eta \cdot \nabla\eta \, dx - \int_{G_H} \eta \mathbf{q} \cdot \nabla\eta \, dx + \int_{\Gamma_A \cup \Gamma_+} \mathbf{q} \cdot \boldsymbol{\nu} \eta^2 \, d\Gamma + \int_{G_H} V\eta^2 \, dx$$

$$\geq D_0 \int_{G_H} \nabla\eta \cdot \nabla\eta \, dx + \frac{1}{2}\int_{\Gamma_A \cup \Gamma_+} \mathbf{q} \cdot \boldsymbol{\nu} \eta^2 \, d\Gamma + V_0 \int_{G_H} \eta^2 \, dx$$

$$+\frac{1}{2}\int_{G_H} LT(y_R - y)\eta^2 \, dx \geq D_0 \int_{G_H} \nabla\eta \cdot \nabla\eta \, dx + V_0 \int_{G_H} \eta^2 \, dx.$$

The theory of linear elliptic equations implies the existence and uniqueness of $\eta \in W^{1,2}(G_H)$ with $\eta = 0$ on $\Gamma_P \cup \Gamma_-$. $\square$

Remark 4.15 Let us note that $\nabla \cdot \mathbf{q} = LT(y_R - y)$ a.e. in G_H is satisfied if the solution of the air flow BVP (4.5) is sufficiently smooth in G_H, which can be guaranteed by sufficient smoothness of the transmissivity T and the boundary Γ. $\square$

Remark 4.16 One can prove theoretically the existence and uniqueness of the solution of the transport problem (with no diffusion, $D = 0$) under some reasonable conditions. We refer the reader for this to the Dautray and Lions [DL93] Chapter XXI. $\square$

Theorem 4.17 (Regularity) Let $D, \rho_R, V, \mathbf{q} \in C^\infty(\overline{G_H})$ such that $D, \rho_R > \delta > 0$. Then $\eta \in C^\infty(G_H)$.

Proof: The proof follows from Dautray, Lions [DL90] (p. 585, Proposition 5). $\square$

4.5 Differentiation of the Transport Equations and the Objective Function

We now want to consider the dependence of the solutions $\mathbf{q}$ and η of problems (3.23), (3.24) on the control variable u. We introduce the derivatives

$$y_k = \frac{\partial y}{\partial u_k}, \quad \mathbf{q}_k = \frac{\partial \mathbf{q}}{\partial u_k}.$$

For the air flow equation, we get the following problem for the derivative with respect to u_k

$$\begin{cases} \nabla \cdot \mathbf{q}_k & = -LTy_k - \delta_k & \text{in } G \\ \mathbf{q}_k & = -T\nabla y_k & \text{in } G \\ \mathbf{q}_k \cdot \boldsymbol{\nu} & = lTy_k + q_{N,k} & \text{on } \Gamma_N \\ y_k & = y_{D,k} & \text{on } \Gamma_D \\ \frac{1}{2\pi R_j} \int_{\Gamma_j} y_k \, d\Gamma & = y_{j,k} & \text{for } \Gamma_j \subset \Gamma_P \end{cases} \quad (4.17)$$

The boundary conditions have to be chosen carefully in accordance with BVP (3.23). By $y_{D,k}, q_{N,k}, y_{j,k}, \gamma_{j,k}$ we denote the derivatives of the boundary conditions $y_D, q_N, y_j,$ and γ_j with respect to the control variable u_k, $k = 1, \ldots, n$. For the flux boundary conditions [F] given in equation (3.3) and assuming that y_D, q_N do not depend on the control variables u_k, we obtain

$$y_{D,k} = 0, \quad q_{N,k} = 0, \quad \gamma_{j,k} = \delta_{jk} \frac{T(s)}{\int_{\Gamma_j} T(\sigma) d\Gamma}, \quad y_{j,k} = 0$$

with the Kronecker symbol δ_{jk}. For the pollutant transport, using $\eta_k = \frac{\partial \eta}{\partial u_k}$ and evaluating $\mathbf{q} \cdot \nabla \eta_k$ and $\mathbf{q}_k \cdot \nabla \eta$ we get

$$\begin{cases} -\nabla \cdot (D\rho_R \nabla \eta_k) + \mathbf{q} \cdot \nabla \eta_k + V\eta_k = -\mathbf{q}_k \cdot \nabla \eta & \text{in } G \\ \mathbf{F}_k = -D\rho_R \nabla \eta_k + \mathbf{f}_k & \text{in } G \\ \mathbf{f}_k = \eta_k \mathbf{q} + \eta \mathbf{q}_k & \text{in } G \\ \mathbf{F}_k \cdot \boldsymbol{\nu} = \mathbf{f}_k \cdot \boldsymbol{\nu} & \text{on } \Gamma_+ \\ \eta_k = 0 & \text{on } \Gamma_- \\ \int_{\Gamma_j} \mathbf{F}_k \cdot \boldsymbol{\nu} \, d\Gamma = 0 & \text{for } \Gamma_j \subset \Gamma_P. \end{cases} \quad (4.18)$$

Now we are able to compute the derivative of the cost functional

$$\frac{\partial J}{\partial u_k} = -c_1 \frac{\partial}{\partial u_k} \sum_{j=1}^{n} \int_{\Gamma_j} \mathbf{F} \cdot \boldsymbol{\nu} \, d\Gamma - c_2 \quad (4.19)$$

$$= -c_1 \sum_{j=1}^{n} \int_{\Gamma_j} \frac{\partial \mathbf{F}}{\partial u_k} \cdot \boldsymbol{\nu} \, d\Gamma - c_2 \quad (4.20)$$

with

$$\frac{\partial \mathbf{F}}{\partial u_k} = -D\rho_R \nabla \eta_k + \frac{\partial}{\partial u_k}(\eta \mathbf{q}) = -D\rho_R \nabla \eta_k + \eta_k \mathbf{q} + \eta \mathbf{q}_k. \quad (4.21)$$

4.5. Differentiation of the Transport Equations

For vanishing pollutant flow through the exterior boundary we can use the Gauss theorem and equation (3.6) and get

$$\frac{\partial J}{\partial u_k} = -c_1 \frac{\partial}{\partial u_k} \sum_{j=1}^{n} \int_{\Gamma_j} \mathbf{F} \cdot \boldsymbol{\nu} \, d\Gamma - c_2 = -c_1 \frac{\partial}{\partial u_k} \int_{G_H} \nabla \cdot \mathbf{F} \, d\mathbf{x} - c_2 \quad (4.22)$$

$$= -c_1 \frac{\partial}{\partial u_k} \int_{G_H} V(\eta_S - \eta) \, d\mathbf{x} - c_2 = c_1 \int_{G_H} V \eta_k \, d\mathbf{x} - c_2 \,. \quad (4.23)$$

Remark 4.18 When the streamline method (see Section 3.2.4) is used for calculating the contaminant transport, the evaluation of $\nabla \eta$ which is needed in formula (4.18) is not directly possible. We refer the reader for numerical algorithm for optimization to Section 7.4. □

Chapter 5

Optimization of Simple Well Configurations

5.1 One Single Active Well

We will now consider the case of a single active well at the origin. We take dimensionless variables as in Section 3.2.1, omitting the hat. We assume that diffusion vanishes, i.e., $D = 0$, the source is radially symmetric ($V(\mathbf{x}) = V(r) \geq 0$) and the transmissivity T is constant.

Without Leakage

From the air flow equation we get the radial component of the air flow

$$q_r = -\frac{u}{2\pi r}.$$

Using the relative mass concentration η, the pollutant equation writes as follows:

$$-\frac{u}{2\pi r}\partial_r \eta = \nabla \eta \cdot \mathbf{q} = V(r)(1-\eta)$$

or with $\tilde{\eta} = 1 - \eta$

$$\frac{\partial_r \tilde{\eta}}{\tilde{\eta}} = \frac{2\pi V(r)}{u} r.$$

Together with the limit condition $\lim_{r \to \infty} \tilde{\eta} = 1$ we get

$$\eta(r) = 1 - \exp\left(-\frac{2\pi}{u} \int_r^\infty r V(r) dr\right). \tag{5.1}$$

5.1.1 Constant Volatilization on a Disk

For constant $V = 1$ on a disk around the origin with radius $r_0 = 1$ (and vanishing outside) we get the radially symmetric solution of equation (5.1)

$$\eta(r) = 1 - e^{-\frac{\pi}{u}(1-r^2)}$$

and the derivatives are

$$\partial_r \eta(r) = -2r\frac{\pi}{u}e^{-\frac{\pi}{u}(1-r^2)} \;;\; \eta_1(r) = \partial_u(r)\eta = -(1-r^2)\frac{\pi}{u^2}e^{-\frac{\pi}{u}(1-r^2)}\;.$$

The last derivative can also be calculated by solving (4.17) and (4.18) which gives the following problem for η_1

$$\begin{cases} q_1 &= \dfrac{1}{2\pi r} & \text{on }]0,1] \\ -u\partial_r \eta_1 + 2\pi r \eta_1 &= \dfrac{2\pi r}{u}e^{-\frac{\pi}{u}(1-r^2)} & \text{on } [0,1] \\ \eta_1(1) &= 0\;. \end{cases} \tag{5.2}$$

The mass extraction rate is

$$J(u) = u\eta(0) = u\left(1 - e^{-\frac{\pi}{u}}\right) \approx \begin{cases} u & \text{for } u \searrow 0 \\ \pi & \text{for } u \to \infty\;. \end{cases}$$

This dependence is shown in Figure 5.1. From this one can conclude that it is a waste of effort to increase the dimensionless total discharge u more than over 20.0 or 30.0.

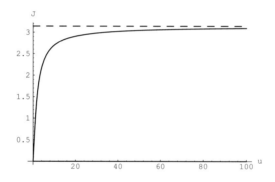

Figure 5.1: Pollutant extraction rate $u\eta(0)$ vs. pumping rate u for a single well with $r_0 = 1$. The dashed line indicates the value π.

5.1.2 Bell-Shaped Volatilization

The soil contamination is not uniform in many real cases. Sometimes, there are only a few "hot spots" containing the chemicals. Such cases could be modeled using a bell-shape volatilization function. A typical example of this is

$$V(r) = e^{-\frac{r^2}{2}}, \tag{5.3}$$

we get

$$\eta(r) = 1 - \exp\left(-\frac{2\pi}{u}\exp\left(-\frac{r^2}{2}\right)\right). \tag{5.4}$$

5.1. One Single Active Well

With Leakage

For vanishing diffusion ($D = 0$), constant leakage L and constant transmissivity T we use the dimensionless form from Section 3.2.1. Then we get for a single active well with discharge u in an infinite domain

$$y(r) = -\frac{u}{2\pi} K_0(\sqrt{L}r)$$

with $r = |\mathbf{x} - \mathbf{x}_1|$ and thus for the radial component of the air flow

$$q_r(r) = -\frac{u\sqrt{L}}{2\pi} K_1(\sqrt{L}r)$$

with asymptotic behavior

$$y(r) \sim -\frac{\pi u}{2}\sqrt{\frac{\pi}{2\sqrt{L}r}} e^{\sqrt{L}r} \tag{5.5}$$

$$q_r(r) \sim -\frac{\pi u\sqrt{L}}{2}\sqrt{\frac{\pi}{2\sqrt{L}r}} e^{\sqrt{L}r} \tag{5.6}$$

for $r \to \infty$.

From (3.48) we get

$$\frac{d\eta}{dt} = V(r)(1 - \eta) + Ly(r)\eta$$

which can be written as

$$\frac{d\eta}{dr} = \frac{V(r)(1 - \eta) + Ly(r)\eta}{q_r(r)}. \tag{5.7}$$

Another possibility is to write this equation with the arc-length s in the r–η-phase-plane as independent variable. Then one gets

$$\frac{dr}{ds} = \frac{q_r(r)}{w}$$

and

$$\frac{d\eta}{ds} = \frac{1}{w}(V(r)(1 - \eta) + Ly(r)\eta)$$

with

$$w = \sqrt{(q_r)^2 + (V(r)(1 - \eta) + Ly(r)\eta)^2}.$$

Numerically obtained solutions for constant volatilization are shown in Figure 5.2. One can see that for large L and small u there is hardly any dependence of the value $\eta(0)$ on the initial value $\eta(1)$. This indicates that in this situation the extraction rate $J(u)$ is practically proportional to u.

Lemma 5.1 *Equation (5.7) has one and only one bounded solution. For this solution one has $0 < \eta(r) < 1$ for all $r > 0$. If V is constant, one gets $\lim_{r \to \infty} \eta(r) = 1$.*

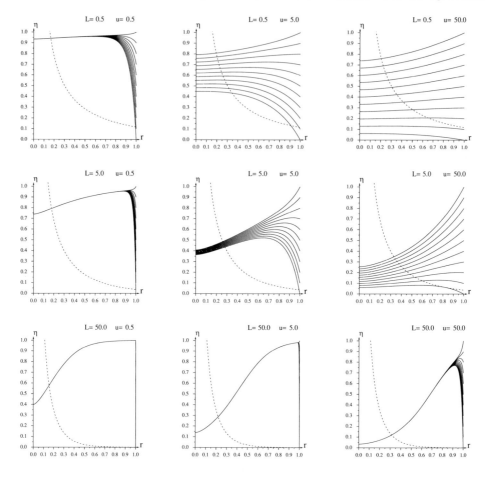

Figure 5.2: Relative concentration η as function of the distance r for varying initial concentrations at $r = 1.0$ and different values of leakage L and discharge u. The black dashed curve indicates the normalized absolute value $|\mathbf{q}|$ of the flow rate, i.e., in the case $u = 1$.

Proof: We write equation (5.7) as

$$\frac{d\eta}{dr} = f(r)\eta - g(r)$$

with positive functions

$$f(r) = \frac{V(r) - Ly(r)}{-q_r(r)} \quad \text{and} \quad g(r) = \frac{V(r)}{-q_r(r)}.$$

First we prove existence. The right-hand side $f(r)\eta - g(r)$ of our differential equation is positive for $\eta = 1$ and negative for $\eta = 0$ for all $r > 0$. Therefore, if a solution satisfies

$$0 \leq \eta(r_1) \leq 1$$

5.1. One Single Active Well

for any $r_1 > 0$, then it satisfies
$$0 \leq \eta(r) \leq 1$$
for all $0 \leq r \leq r_1$. We obtain a bounded solution by defining η_{r_1} as the solution for which $\eta_{r_1}(r_1) = \eta^*$ with arbitrary $0 \leq \eta^* \leq 1$ holds and then constructing the limit
$$\eta_\infty(r) = \lim_{r_1 \to \infty} \eta_{r_1}(r)$$
for all $r \geq 0$.

Now we prove uniqueness. Let $r_0 > 0$ be fixed. Then any solution of the homogeneous equation
$$\frac{d\eta}{dr} = f(r)\eta$$
is a multiple of
$$\eta(r) = e^{F(r)}$$
where
$$F(r) = \int_{r_0}^{r} f(s)ds.$$
Form the "variations of constants" we get for any inhomogeneous solution
$$\eta(r) = e^{F(r)} \left(\eta(r_0) - \int_{r_0}^{r} e^{-F(s)} g(s) ds \right).$$

Now we know from the asymptotic expansions of the Bessel functions that $\frac{y(r)}{q_r(r)} \to \frac{1}{\sqrt{L}}$ and thus $F(r) \to \infty$ for $r \to \infty$, taking into account that $V \geq 0$. Therefore, there is one and only one $\eta(r_0)$ for which $\eta(r)$ remains bounded for large r, namely
$$\eta(r_0) = \int_{r_0}^{\infty} e^{-F(s)} g(s) ds.$$

We have shown that there is a unique solution having the desired properties.

We now assume that $V > 0$ is constant, i.e., $V = 1$ in dimensionless variables. Then the right-hand side of our differential equation vanishes along the equilibrium curve E
$$\eta_E(r) = \frac{1}{1 - Ly(r)},$$
see Figure 5.3. In the same way as above, we can see that for a solution with
$$\eta_E(r_1) \leq \eta(r_1) \leq 1$$
for an arbitrary r_1, we also have
$$\eta_E(r) \leq \eta(r) \leq 1$$

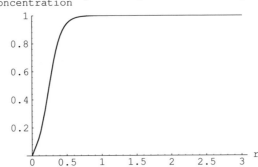

Figure 5.3: Equilibrium curve $\eta_E(r)$ for $u = 1$ and $L = 100$.

for all $0 \leq r \leq r_1$. Since $\eta_E(r_1)$ is a strictly increasing function with $\lim_{r \to \infty} \eta_E(r) = 1$, we get

$$\lim_{r \to \infty} \eta_\infty(r) = 1 \;,$$

which completes the proof. □

Remark 5.2 With the notation from the previous proof one gets

$$J(u) = u \lim_{r \searrow 0} e^{F(r)} \int_0^\infty e^{-F(s)} g(s) ds$$

for the extraction rate. □

Figure 5.4 shows the extraction rate.

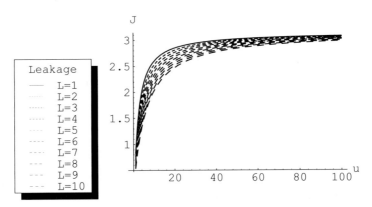

Figure 5.4: Extraction rate from a single well as function of the discharge u for varying leakage term $L = 0.0, 1.0, \ldots, 10.0$. The initial concentration at $r = 1.0$ is zero.

5.1. One Single Active Well

Bell-Shaped Volatilization

Now we assume that the leakage L is constant and the volatilization is bell shaped. In particular, we assume that we have

$$V(r) = \exp(-\frac{r^2}{2}).$$

Results solving system (3.48) are shown in Figure 5.5.

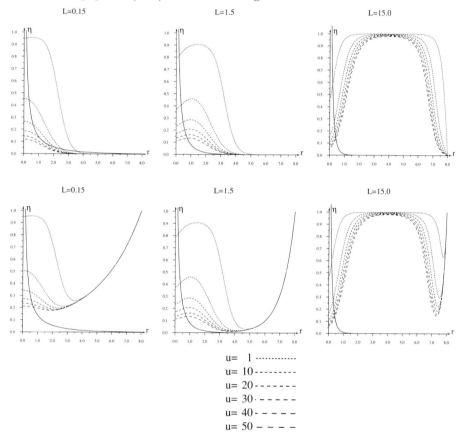

Figure 5.5: Relative concentration η as function of the distance r for a bell-shaped volatilization function $V(r)$ (black solid curve) and different values of leakage L and discharge u. The black dashed curve indicates the normalized absolute value $|\mathbf{q}|$ of the flow rate, i.e., in the case $u = 1$.

Figure 5.6 shows the extraction rate.

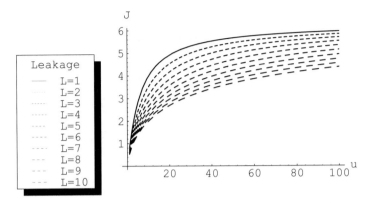

Figure 5.6: Extraction rate from a single well as function of the discharge u for varying leakage term $L = 1.0, \ldots, 10.0$. The volatilization is assumed to be bell-shaped. The initial concentration at $r = 8.0$ is zero.

5.2 Two Active Wells

With No Leakage

We consider the case of two active and one passive well, i.e. $n = 2$, $N = 3$, $C_{sat} = 1$ and $u_3 = -1$, $0 \leq u_1 \leq 1$, $u_2 = 1 - u_1$. In equation (3.6) we take

$$V(\mathbf{x}) = \begin{cases} 0.1 & \text{for } |\mathbf{x} - \mathbf{x}_1| \leq 2 \\ 0.2 & \text{for } |\mathbf{x} - \mathbf{x}_2| \leq 1 \\ 0 & \text{else .} \end{cases}$$

Diffusion is neglected. Figure 5.7 shows the total pollutant extraction rate (i.e., the sum of the extraction rates of both active wells) depending on u_1 with an optimum at $u_1 \approx 0.7$. For this value, the streamlines of the underlying flow field are shown in Fig .5.8.

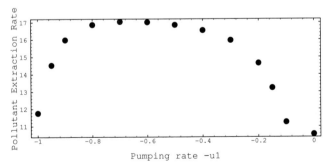

Figure 5.7: Contaminant extraction rate vs. pumping rate u_1.

5.3. Configurations with a Few Wells

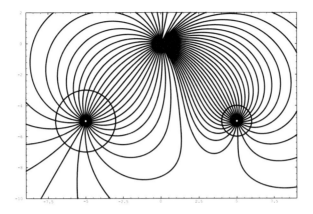

Figure 5.8: Streamlines for the optimal pumping rate $u_1 = 0.7$.

5.3 Configurations with a Few Wells

Bell-Shaped Volatilization

We again assume a volatilization like in Section 5.1.2, but now we have a given number of active wells at our disposal. We consider symmetric configurations of the wells. Our goal is to find optimal configurations of the wells for a given set of parameters u, L. First, we consider a set of m wells, $m \in \{2, 3, 4, 6\}$ with same discharge uniformly distributed over a circle of radius r_0 around the pollution center. The total discharge of all these wells is assumed to be u. We also consider the limit case $m \to \infty$, i.e., a circle of discharge density $\frac{u_r}{2\pi r_0}$, together with a well at the origin of the co-ordinate system with discharge $u_0 = u - u_r$. Since we then have a radially symmetric problem, explicit expressions for y can be derived in the same way as in Section 3.2.3, see (3.40). We get

$$y(r) = \begin{cases} C_I I_0(\sqrt{L}r) + C_0 K_0(\sqrt{L}r) & \text{for } r < r_0 \\ (C_K + C_0) K_0(\sqrt{L}r) & \text{for } r \geq r_0 \end{cases} \quad (5.8)$$

$$q(r) = \begin{cases} -C_I \sqrt{L} I_1(\sqrt{L}r) + C_0 \sqrt{L} K_1(\sqrt{L}r) & \text{for } r < r_0 \\ (C_K + C_0) \sqrt{L} K_1(\sqrt{L}r) & \text{for } r \geq r_0 \end{cases}.$$

The coefficients are determined by the continuity condition for y at r_0, i.e., $C_I I_0(\sqrt{L} r_0) = C_K K_0(\sqrt{L} r_0)$ and the discharge condition $\lim_{\delta r \to 0} q(r_0 - \delta r) - q(r_0 + \delta r) = \frac{u_r}{2\pi r_0}$. From this we get

$$C_I = -\frac{u}{2\pi \sqrt{L} r_0} \frac{K_0(\sqrt{L} r_0)}{I_0(\sqrt{L} r_0) K_1(\sqrt{L} r_0) + I_1(\sqrt{L} r_0) K_0(\sqrt{L} r_0)}$$

$$C_K = \frac{I_0(\sqrt{L} r_0)}{K_0(\sqrt{L} r_0)} C_I \,, \quad C_0 = -\frac{u - u_r}{2\pi} \,. \quad (5.9)$$

In Figure 5.9 the negative of the square of pressure y and the radial flow q are displayed for $u_0 = 0$, $r_0 = 0.5$, discharge density $\frac{1}{2\pi r_0}$ and leakage $L = 15$.

For some parameter combinations there exists a radius $r' \leq r_0$ where the flux vanishes

$$C_0 K_1(\sqrt{L}r') = C_I I_1(\sqrt{L}r') . \qquad (5.10)$$

The restriction $r' \leq r_0$ is clear because C_0 and C_K have both negative signs and thus the airflow will always be directed to the origin outside the ring-shaped well. Since I_1 is strictly increasing and K_1 strictly decreasing, equation (5.10) has a solution r' if and only if

$$C_0 K_1(z_0) \geq C_I I_1(z_0) \qquad (5.11)$$

where we have introduced the abbreviation $z_0 = \sqrt{L} r_0$. Remember that $C_0, C_I \leq 0$. Using the formulas (5.9) we get the equivalent condition

$$\left(1 - \frac{u_r}{u}\right) \leq z_0 \frac{I_1(z_0)}{K_1(z_0)} \frac{I_0(z_0)K_1(z_0) + I_1(z_0)K_0(z_0)}{K_0(z_0)}$$
$$= z_0 I_1(z_0) \left(\frac{I_0(z_0)}{K_0(z_0)} + \frac{I_1(z_0)}{K_1(z_0)}\right) . \qquad (5.12)$$

We can see that for the l.h.s. $0 \leq 1 - \frac{u_r}{u} \leq 1$ holds whereas the r.h.s. is strictly increasing and unbounded, vanishing for $z_0 = 0$. This means that there is a maximal z_0' below which condition (5.12) is violated for sufficiently small u_r, uniquely determined by

$$z_0' I_1(z_0') \left(\frac{I_0(z_0')}{K_0(z_0')} + \frac{I_1(z_0')}{K_1(z_0')}\right) = 1 , \text{ the solution is } z_0' = 0.824451 .$$

The conclusion from (5.12) is that for $z_0 < z_0'$ and sufficiently small u_r:

$$\frac{u_r}{u} < 1 - z_0 I_1(z_0) \left(\frac{I_0(z_0)}{K_0(z_0)} + \frac{I_1(z_0)}{K_1(z_0)}\right) ,$$

the flow field is dominated by the central well and has negative values everywhere. If $z_0 \geq z_0'$, this is never the case. The results are shown in Figures 5.10, 5.12 and 5.11.

First we discuss Figure 5.10. Let us fix the number of wells for a moment. Then the extraction rate is increasing with increasing discharge independently of the well configuration. The strategic position of the centered well (look "ring+well") is dominant for large discharges. On the other hand, it makes no sense to pump very strongly when the leakage L is small (cf. case $u = 40$, $L = 0.15$). Here the extraction rate is decreasing with an increasing number of wells. The reason is, the inflow of a clean air through the surface (which is modeled by the leakage term L in the 2-D vertically integrated model) is not sufficient for large discharges and thus the contaminated area is not cleaned effectively. Let us note then for $L = 0$ there will be an almost "dead" region around the center of pollution, which will be larger for large discharges.

Now we discuss Figure 5.12. Here the leakage L is 100-times larger than in Figure 5.10. We can see that the more wells we use for a given discharge, the larger the extraction rate is. This is the main difference to Figure 5.10.

The Figure 5.11 shows the behavior of the total extraction rate for a constant discharge $u = 1$ and different well configurations. The extraction rate decreases with increasing L.

5.3. Configurations with a Few Wells

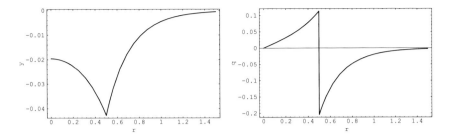

Figure 5.9: Square of pressure and radial flow for a ring sink of radius $r_0 = 0.5$ with constant discharge density and leakage $L = 15$.

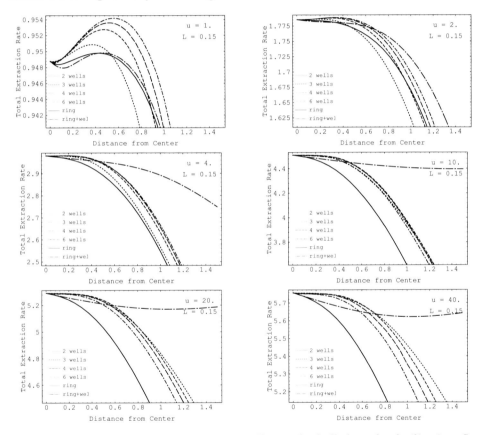

Figure 5.10: Total extraction rates for the pollutant for bell-shaped volatilization, $L = 0.15$ and different values for the discharge.

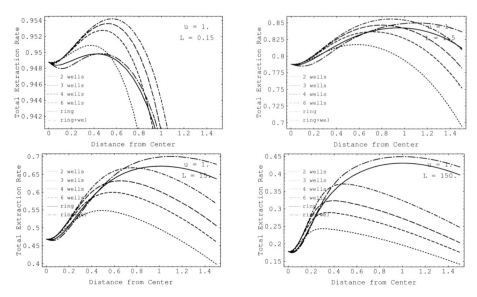

Figure 5.11: Total extraction rates for the pollutant for bell-shaped volatilization, $u = 1$ and different values of leakage.

5.3. Configurations with a Few Wells

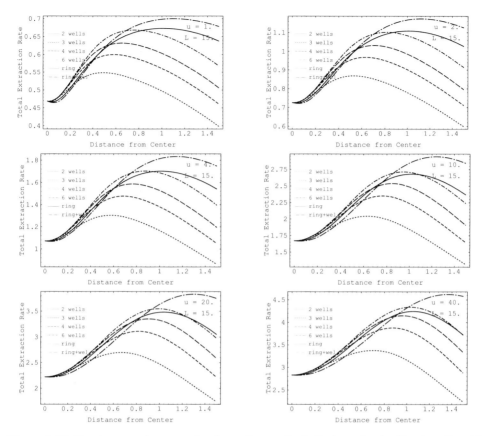

Figure 5.12: Total extraction rates for the pollutant for bell-shaped volatilization, $L = 15$ and different values for the discharge.

Chapter 6

Estimating the Coefficients

6.1 Pedotransfer Model and Parameter Estimation Strategy

In most practical applications of soil venting directly measured model parameters are not available. However, in many cases, qualitative descriptions, such as soil protocols from boreholes, and soil survey data may be available. The following pedotransfer model and strategy to estimate model parameters from qualitative descriptions has been used and is proposed for application to contaminated sites in which the porous soil matrix consists predominantly of coarse-textured, sandy material.

Basic Assumptions

1. Only the water-unsaturated upper parts of the soil are considered.

2. The water table is assumed to be constant in time.

3. The soil material can be classified into three groups: A) originally layered, B) refilled construction sand, and C) rubble material. In all three groups, sandy materials are the dominating volumetric portion of the porous matrix. Qualitatively described differences in soil bulk density as well as secondary or minor components, such as rock and wood debris, can be used to modify the estimates of porosity and permeability.

4. The parameter estimation procedure is based on techniques used in geohydraulic and soil physics (see Busch and Luckner [BL73]).

The estimation procedure is based on the following steps:

6.1.1 Grouping of Soil Substrates

The qualitative analysis of borehole soil protocols allows a grouping of the soil materials in three classes having different soil components, origins, degrees of anthropogenetic disturbance, and structure. For each of the groups the estimation procedure has to be different. For example, by comparing about 100 soil protocols of the test site of Kertess we find the following classes

Originally layered materials
They are fluviatile deposits of banded or finely stratified coarse to fine sands in its originally layered structure.

Refilled materials
They are predominately sandy soil materials for house and road constructions that have been intensely mixed and relocated and which mostly contain some stones, debris and rubble material (up to 20% stone content). Refilled sands show no easy to determine structures; the refilled sand may be either compacted or highly porous, depending on the treatment and technique used during dumping.

Rubble material originating from building or road construction
The rubble can be a highly disordered mixture of sand and stones or rock and wood debris. The rubble material is highly heterogeneous and is defined for a stone content of around 50%. It may contain fractures and fissures.

6.1.2 Grouping of the Texture of the Fine Soil Material (equivalent grain diameter smaller than $2\ mm$)

The fine soil material ($< 2\ mm$) for each substrate group is further divided into three texture classes: coarse, medium, and fine sand. The grouping will be done according to the major texture attribute given in the soil protocol. A further subdivision of the three texture classes is done according to the kind and proportion of secondary and minor components or depending on the estimated soil bulk density that has been observed and described in the protocols.

6.1.3 Estimation of the Cumulative Particle-Size Distribution

The qualitative description of the soil materials from borehole protocols are transformed into relative proportions of grain-size fractions using the DIN 4022 [DIN87].

Table 6.1 shows average values of the particle-size distribution estimates for originally layered, undisturbed sands obtained from selected borehole protocols. The particle-size distributions in Table 6.1 are used also for estimating the properties of the fine soil matrix of the refilled and rubble substrates. The grain-size classes in Table 6.1 consist of clay T ($< 0.002\ mm$), silt U ($0.002 - 0.063\ mm$), fine sand fS ($0.063 - 0.2\ mm$), medium sand mS ($0.2 - 0.63\ mm$), and coarse sand gS ($0.63 - 2\ mm$).

6.1.4 Graphical Interpretation of the Particle-Size Distribution

For graphical interpretation of the particle-size distribution (Table 6.1) a cumulative grain-size distribution function in semi-logarithmic form (see Figure 6.1) based on the

6.1. Pedotransfer Model

number	soil material abbreviations	cumulative frequency in texture classes of max. particle diameter [mm]				
		T 0.002	U 0.063	fS 0.2	mS 0.63	gS 2
1	coarse sand 1: gS, ms	0	0	5	25	100
2	coarse sand 2: $gS, \overline{ms}$	0	3	8	45	100
3	medium sand 1: mS, fs, u'	2	10	45	95	100
4	medium sand 2: mS, fs, gs, u	2	15	35	85	100
5	fine sand 1: $fS, ms1$	3	12	65	85	100
6	fine sand 2: $fS, ms2$	2	5	65	90	100

Table 6.1: Kertess: Estimation of the cumulative particle-size distribution for six representative soil materials (for Kertess) obtained from qualitative soil descriptions (borehole soil protocols) according to DIN 4022. The texture-classes are defined as T for clay, U for silt, and fS is fine, mS is medium, and gS denotes coarse sand (the large symbol denotes the main texture component). The six materials are coarse, medium and fine sands which contain differing amounts of secondary components, such as medium – ms, fine – fs, and coarse – gs sandy material, as well as silty – u material (overstrike denotes a relatively large and $'$ a relatively small portion of the secondary component, and $ms1$ and $ms2$ describe two different types of medium sandy materials as secondary components of the fine sand.

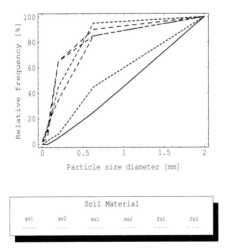

Figure 6.1: Kertess: Originally layered sandy soils.

gravimetric portions is used. (An analytical grain-size distribution function will be used below). By log-linearly interpolating the relative cumulative values between the estimates the following coefficients for parameter estimation can be derived: The degree

soil material	d_{10} [mm]	d_{60} [mm]	U [-]	porosity ε $\left[\frac{m^3}{m^3}\right]$			bulk density ρ_b $\left[\frac{kg}{m^3}\right]$		
				loose	medium	dense	loose	medium	dense
1	0.260	1.07	4.12	0.37	0.32	0.29	1.66	1.80	1.87
2	0.210	0.87	4.14	0.36	0.32	0.29	1.69	1.80	1.87
3	0.060	0.28	4.67	0.35	0.31	0.28	1.72	1.83	1.91
4	0.016	0.36	22.50	0.32	0.27	0.25	1.80	1.92	1.99
5	0.069	0.18	2.60	0.38	0.34	0.32	1.64	1.74	1.80
6	0.027	0.18	6.67	0.34	0.30	0.27	1.74	1.86	1.93

Table 6.2: Estimates of the porosity and bulk density for the originally layered sandy materials and for loose, medium, and dense packings of the particles. The terms d_{10} and d_{60} denote the particle-size at 10% and 60% of the cumulative particle-size distribution function, respectively, and U is the degree of non uniformity.

of non uniformity $U = \frac{d_{60}}{d_{10}}[L/L]$, is the relation between the effective particle-size diameter at a relative cumulative portion of 60% d_{60}, and 10% d_{10}, and related to the gravimetrically-based semi-logarithmically drawn cumulative grain-size distribution function. The porosity $n[L^3/L^3]$, is defined as

$$n = \frac{V_t - V_f}{V_t}, \qquad (6.1)$$

where V_t is the total volume of the solid phase $[L^3]$. Estimations of the porosity are based on qualitative estimates of the soil bulk density classified as loose, medium, and dense. The porosity can be estimated after Busch and Luckner [BL73] Fig. 3/31 from the degree of non uniformity U. Table 6.2 shows the estimated values for the grain-size data presented in Table 6.1 including the values of the soil bulk density ρ_b $[ML^{-3}]$. The bulk density is defined as $\rho_b = (1-n)\rho_s$ $[ML^{-3}]$, where the density of the solid phase of the soil, $\rho_s = 2,65\frac{g}{cm^3}$, is assumed to be the value for quartz sand.

6.1.5 Estimation of the Hydraulic Conductivity at Pore Water Saturation

The Kozeny-Carman equation is used for estimating the specific permeability K $[L^2]$ of non-binding, sandy-gravel porous materials expressed in terms of Busch and Luckner [BL73]

$$K = \frac{d_w^2}{72C^2T^*}\frac{(n')^3}{(1-n)^2}, \qquad (6.2)$$

where d_w is the effective hydraulic grain size [-], T^* denotes the tortuosity of the pore system [-], n is the total and n' the flow effective porosity [-]. The permeability coefficient k $[LT^{-1}]$ is related to K by

$$k = K\frac{g}{\nu}, \qquad (6.3)$$

6.1. Pedotransfer Model

soil mat.	$C^* * 10^{-3}$			K_w $*10^{-12}$ $[m^2]$ medium	k_w $*10^{-5}$ $[\frac{m}{s}]$ medium	k_w $[\frac{cm}{day}]$ medium	range from	to	residual water saturation $s_{ob} = \frac{nb}{n}[-]$
	loose	medium	dense						
1	1.6	1.20	0.95	81.120	60.747	5248.6	4155	6997	0.02
2	1.6	1.20	0.95	52.900	39.614	3422.7	2710	4564	0.05
3	1.5	1.15	0.92	4.140	3.100	262.3	210	342	0.25
4	1.2	0.85	0.65	0.218	0.163	13.8	11	20	0.60
5	1.7	1.35	1.10	6.430	4.810	407.2	332	512	0.25
6	1.4	1.10	0.85	0.802	0.601	51.9	40	66	0.45

Table 6.3: Estimates of the specific permeability K_w, using Hazen's approximation, the hydraulic conductivity coefficient k_w, and the residual water saturation s_{ob}, of the six sandy soil materials for the originally layered material. The term C^* is an empirical proportionality factor.

where g $[LT^{-2}]$ is the gravity and ν is the temperature dependent kinematic viscosity $[L^2T^{-1}]$. We first estimate the permeability using a simplified form of (6.2) (Hazen-approximation)

$$K = C^* d_{10}^2, \tag{6.4}$$

where d_{10} is the grain size at a value of 10 % of the cumulative grain-size distribution function $[L]$ and C^* is a factor derived from empirical relations of the degree of non uniformity U (cf. Busch and Luckner [BL73] Fig. 3/32).

Assuming an average soil temperature of $10^\circ C$, we obtain the permeability coefficient $k_{10^\circ C} = C^{**}_{10^\circ C} d_{10}^2$, with $C^{**} = C^* \frac{g}{\nu_{10^\circ C}}$. The proportionality factor C^{**} was found by Hazen to be $C^{**} = 11600 \frac{m^{-1}}{s^{-1}}$ for sand having $U < 5$. Busch and Luckner [BL73] Fig. 3/32 suggest C^{**} to be a function of U. For oil, air, and water at certain temperatures of the system permeability coefficients are derived, e.g., from

$$k = k_{10^\circ C} \frac{\nu_{10^\circ C}}{\nu}. \tag{6.5}$$

The kinematic viscosity for water at $10^\circ C$, $\nu_{10^\circ C,w}$, is given in Busch and Luckner as $1.31 10^{-6} \frac{m^2}{s}$ and for Air, $\nu_{10^\circ C,l}$, with $1.42 10^{-2} \frac{m^2}{s}$.

In the air-water system, the residual water saturation, sob, or the non-moving portion of the wetting fluid for the case air replaces water can be described as a function of the permeability coefficient for water (Table 6.3). The relative saturation $s_{ob} = \frac{n_b}{n}$ is the portion of the porosity filled with the wetting fluid (here: water) of the total porosity. The specific permeability of the porous medium for water and air, K, the residual saturation of the pore space with water, s_{ow}, NAPL-fluid (contaminant), s_{oc}, and air (,e.g., a discontinuous gas phase) s_{ol} is considered by additionally reducing the phase-filled porosity by the immobile portion of the liquid phase to a value of the flow-effective porosity for each phase by:

$$n' = n(1 - s_{ob} - s_{oc} - s_{ol}). \tag{6.6}$$

soil material	d_w [10^{-3}m]	C [-]	T^* [-]	s_{ow} [-]	s_{oc} [-]	n' [-]	n [-]	K_w [$10^{-12}m^2$]
1	0.533	1.05	2.0	0.04	0.01	0.304	0.32	111.00
2	0.431	1.05	2.0	0.05	0.02	0.298	0.32	66.54
3	0.129	1.10	2.2	0.15	0.05	0.248	0.31	2.78
4	0.048	1.05	2.2	0.30	0.10	0.162	0.27	0.11
5	0.124	1.10	2.1	0.15	0.05	0.272	0.34	3.90
6	0.062	1.10	2.1	0.22	0.08	0.210	0.30	0.40

Table 6.4: Estimates of the specific permeability according to Kozeny-Carman equation for the originally layered material.

Results for the six example soil texture classes are given in Table 6.4.

While estimates of the effective porosity are based on the Hazen-approximate permeability approximation, the porosity values are used to estimate the specific permeability of each soil profile and horizon property using the Kozeny-Carman (6.2). Results of the parameters and estimated permeabilities are given in Table 6.4 and compared to those values calculated by Hazen's approximation. The values of K_w in Table 6.4 are corresponding roughly to those calculated with (6.4) (Table 6.2).

The effective hydraulic grain-diameter d_w is obtained by

$$d_w = d_w(U, d_{10}) \qquad (6.7)$$

from an empirical relation of Beyer (in Busch and Luckner [BL73] Tab. 3/3) using the values given in Table 6.2. The parameter d_w is highly sensitive. The less sensitive form parameter C is estimated according to Köster. The value of the tortuosity factor T^* ranges between 1.5. and 2.5; we use here an average value for homogeneous spherical beds given by Carman with $T^* = 2$. We increased T^* for porous substrates having relatively high bulk densities and inhomogeneities of the grain-size distribution and decrease the value of T^* for materials with relatively low bulk density and homogeneous grain size distributions. Estimates of the flow-effective porosity are based on values of residual saturations of Ferris (in Busch and Luckner [BL73] Tab. 3/34), (s_{ob} for water see Table 6.3), where the values of Ferris have been decreased by one-half of the values of Ferris. The residual saturation of NAPL-fluid, s_{oc}, was assumed to be $\frac{1}{3}$ of that of water. The residual saturation for air, soil, has been neglected in this case assuming that the water saturation of the pore space is low and not approaching saturation; which means all air is participating in transport.

6.1.6 Permeability of the Refilled and Rubble Material

The approach leading to results in Table 6.4 using (6.2) is also used for the estimation of permeabilities of the refilled and rubble material. For refilled material we assume that it can be packed either relatively loosely or densely with densities of the fine soil

6.1. Pedotransfer Model

(< 2 mm) of about $1.5 \frac{g}{cm^3}$. Refilled sands can also be very densely packed depending on the technique used for treating the sand. Estimates of the porosity by the soil bulk density using the relation

$$n = 1 - \frac{\rho_b}{\rho_f} \quad (6.8)$$

have been weighted in Table 6.2 around the average value of $1.5\frac{g}{cm^3}$ (according to the range of values in Table 6.2).

Refilled sands are mostly less structured compared to original layered materials. We assume that the stone- and gravel content will increase the tortuosity compared to the average value of $T^* = 2.25$ for the refilled and of $T^* = 2.5$ for the rubble material. In addition, the values of T^* are modified proportional to the differences of the coarse, medium, and fine sands according to the values given in Table 6.2. For refilled and rubble material, also the effective hydraulic grain radius d_w, will become different compared to the d_w-values of the originally layered materials. Depending on the content of stones and gravel, the relative shift in the grain-size-distribution function is modifying the degree of non uniformity and the d_{10}-value which are used for estimating d_w.

6.1.7 Analytical Interpretation of the Particle-Size Distribution

According to Haverkamp and Parlange [HP86], a van Genuchten-type function (cf. van Genuchten [van80]) can be used for the analytical description of the cumulative particle-size distribution $F(d)$

$$F = \left[1 + (d_g d^{-1})^n\right]^{-m}, \quad (6.9)$$

where d $[L]$ is the particle diameter, d_g $[L]$ is a constant (reference diameter), and n and $m = 1 - \frac{1}{n}$ are empirical parameters. For the optimization, the parameters d_g and m are fitted to the borehole protocol estimates of the grain-size distribution, using the nonlinear curve-fitting program RETC (van Genuchten et al. [vGLY91]). The stone and gravel content described in the soil protocols is assumed to be represented by a particular gravel fraction of a particle diameter between 2 and 63 mm. The grain diameter as a function of the relative portion of the cumulative particle-size distribution function can be calculated using the inverse of (6.9)

$$d = d_g \left[\left(\frac{1}{F}\right)^{\frac{1}{m}} - 1\right]^{m-1}. \quad (6.10)$$

Results of the optimized parameters of (6.9) for the originally layered sands as well as for the coarse sands of the refilled and rubble material are given in Table 6.5. For the coarse sands we assumed an average gravel content of 5% (A), 20% (B), and 50% (C). Using the unimodal function of (6.9) is only possible for coarse textured materials or where we have relatively low gravel content. For materials with a matrix consisting of fine and medium sand the particle-size distribution becomes a bi- or multi-modal shape which cannot be described by (6.9). Since no further information is available about the particle-size distribution of the gravel and stones itself we propose the following simple log-linear grain-size distribution assuming a log-linear relation of $F(d)$ for $2 < d < 63\ mm$

$$F = [1 + (d_g d^{-1})^n]^{-m} \quad 0 < d \leq 2\ mm \\ F = a\log(d) + b \quad 2 < d \leq 63\ mm, \tag{6.11}$$

with the regression coefficients $a_B = 0.1335, b_B = 0.16$ and $a_C = 0.33371, b_C = 0.4$, where index B and C denotes the refilled and rubble material, respectively.

For determining the effective grain diameter according to Beyer (Table 3/3 in Busch and Luckner [BL73]) we consider the particle-size distribution of the gravel fraction explicitly only for the coarse sands. We think that the derivation of the degree of non uniformity, U, from the cumulative particle-size distribution is only valid for materials having a unimodal and log-linear function in the range between d_{10} and d_{60}. For all other porous materials (medium and fine sands) we assume that the effective grain diameter only depends on the particle-size distribution of the fine soil ($< 2\ mm$), which means it is identical to the values of d_w of the original layered material in Table 6.4. The permeability of the refilled and rubble material is then calculated depending on stone-content corrected porosity and tortuosity modification assuming that the stone and rubble component has minor effect on the pore size distribution of the fine material itself. However, the assumption may be questionable, since in porous materials with relatively high stone and gravel components often a dual-porosity system may exist. Besides the sand matrix pore system, rubble materials may show a second, larger pore system which may be formed by pores along the matrix stone surfaces which if continuously may affect the air permeability dominantly. Since there is no information available with respect to the problem of either bimodal grain or pore size distributions we assume unimodal homogeneous matrix properties and flow of air and water only through the matrix.

The porosity n is reduced by stone and rubble material according to

$$n = n^* \left(1 - \frac{V_s}{V_t}\right), \tag{6.12}$$

where n^* is the porosity related to the fine soil ($d < 2\ mm$) calculated using (6.8), and the soil bulk density V_s is the volume of stones ($d > 2\ mm$) estimated by soil borehole protocols to be 5, 20 and 50 %, and V_t is the total volume of the bulk soil. We assume that the soil borehole protocol information regarding the stone components can be interpreted in terms of volume proportions. In general, the amount of minor components is defined by the soil descriptions in terms of "low" is defined to have 0-15%, "medium" as 15-30%, and "high" as 30-40%. We assume a stone content for refilled material of 10-30% and for rubble of 40-60%. Since the stone content affects the particle-size distribution also the values of the coefficients d_{10}, d_{60}, U, and d_w will be modified. The values for the residual saturations are changed according to the changes in the hydraulic permeability introduced by the stone content. However, we will first use the values given in Table 6.4.

6.1.8 Residual Water Saturation

Another problem is the estimation of the residual saturation of the porous medium by water, NAPL or air. The term s_{ob} denotes the saturation of the wetting fluid, while

6.1. Pedotransfer Model

s_{onb} is the saturation of the non-wetting fluids. Using Schweiger's empirical relation, the permeability has to be known in advance to estimate the residual saturation. However, if we have steady-state flow conditions we may neglect dynamic effects, hysteresis, and variability.

6.1.9 Hydraulic Parameter Functions

To describe the hydraulic parameters we use the models of van Genuchten and Mualem (cf. van Genuchten [van80]) (VGM-model)

$$\theta(h) = \theta_r + \frac{\theta_s - \theta_r}{(1 + |\alpha h|^n)^m}, \tag{6.13}$$

$$K(S_e) = K_s S_e^l \left[1 - \left(1 - S_e^{\frac{1}{m}}\right)^m\right]^2 \quad m = 1 - \frac{1}{n}; \quad n > 1, \tag{6.14}$$

and the model of Brooks and Corey (BC-model)

$$\begin{aligned} S_e &= (\alpha h)^{-\lambda}, & \alpha h &> 1 \\ S_e &= 1, & \alpha h &\leq 1, \end{aligned} \tag{6.15}$$

$$K(S_e) = K_s S_e^{l+2+\frac{2}{\lambda}}, \tag{6.16}$$

where θ_r is the residual and θ_s the saturated water content, h is pressure head, K_s is the saturated hydraulic conductivity, $S_e = (\theta - \theta_r)/(\theta_s - \theta_r)$ is the effective saturation, α, n_{VG}, l and λ are empirical parameters. For $n_{VG} \to \infty$ and $n_{VG}m$ remains finite the VGM approaches the BC-relations-ship.

6.1.10 Relative Permeability, Saturation-Pressure, and Relative Permeability Relations

The relative permeability, index r, is defined as

$$K_r = k_r = \frac{K}{K_s}. \tag{6.17}$$

Schweiger (cf. Busch and Luckner [BL73] proposed for sandy soils and 3-phase system air water solid phase the following empirical relation for the relative water permeability

$$\begin{aligned} K_{r,w} &= 0, & 0 &\leq s_w \leq s_{ow}, \\ K_{r,w} &= \left(\frac{s_w - s_{ow}}{1 - s_{ow}}\right)^3 & s_{ow} &< s_w \leq 1 \end{aligned} \tag{6.18}$$

and for air permeability

$$\begin{aligned} K_{r,l} &= 1 & s_{ol} &\leq s_l \leq 1, \\ K_{r,l} &= \left(\frac{s_l}{1 - s_{ol}}\right)^{2.4} & 0 &\leq s_l < s_{ol}. \end{aligned} \tag{6.19}$$

In case the parameters of the VGM- model are available, the approach of Parker et al. [PLK87] can be used to describe the parameters in two- and three-phase systems based on the analysis of the hydraulic parameters of the single-phase water flow system. The two-phase systems saturation-capillary pressure relation of Parker et al. [PLK87] is defined as

$$\begin{aligned} S_j^{ij} &= S_m + (1 - S_m)\left[1 + (\alpha_{ij}h_{ij})^n\right]^{-m}, & h_{ij} &> 0, \\ S_j^{ij} &= 1, & h_{ij} &\leq 0, \end{aligned} \qquad (6.20)$$

where $m = 1 - 1/n$, $i, j = a(air), o(oil), w(water)$, $i \neq j$, $\alpha_{ij} = \alpha\beta_{ij}$; S_m is the minimal or residual saturation of the wetting fluid. Scaling of the shape of the curve to that of the air-water system, using $\beta_{aw} = 1$, allows to derive parameters for the other pairs of liquid phase fluids (ao and ow). Parker et al. [PLK87] suggest for $\beta_{ao} = 1.904$ and $\beta_{ow} = 2.12$. The relative permeabilities as function of the phase saturation are described for water by

$$k_{rw} = \overline{S}_w^l \left[1 - \left(1 - \overline{S}_w^{\frac{1}{m}}\right)^m\right]^2 \qquad (6.21)$$

with

$$\overline{S}_w^{ij} = \frac{S_j^{ij} - S_m}{1 - S_m}, \qquad (6.22)$$

and for air by

$$k_{ra} = C\overline{S}_a^l \left(1 - \overline{S}_t^{\frac{1}{m}}\right)^{2m}, \qquad (6.23)$$

in which $C = 1 + \frac{\kappa}{h_a}$ is included to consider the Klinkenberg effect; $S_t = 1 - S_a$ is the total pore saturation by all fluid phases. For the NAPL phase assuming an average wettability the relative permeability is

$$k_{ro} = (\overline{S}_t - \overline{S}_w)^l \left[\left(1 - \overline{S}_w^{\frac{1}{m}}\right)^m - \left(1 - \overline{S}_t^{\frac{1}{m}}\right)^m\right]^2, \qquad (6.24)$$

where a value of $l = 0.5$ is often assumed. We neglect hysteresis of the pressure-saturation relation in our study, because the mostly sealed surface and the steady-state flow conditions of our test cases provide conditions in which rapid changes between desorption and adsorption are unlikely to occur.

6.1.11 Pedotransfer Functions, Pressure-Saturation Relations

For sandy soils without any organic material the approach of Haverkamp and Parlange [HP86] can be used to estimate the pressure-saturation relation. The approach requires only information of the particle-size distribution and the soil bulk density. The approach is based on the simplified capillary-rise equation in which a packing index γ is included as

$$h = 0.149\frac{\gamma}{d} \qquad (6.25)$$

and assumes that the saturation curve is proportional to the particle-size distribution as

$$\theta = \theta_s F(d), \qquad (6.26)$$

6.1. Pedotransfer Model

	soil material	gravel [%]	d_g [mm]	n [-]	m [-]	λ [-]	d_{10} [mm]	d_{60} [mm]	d_w/d_{10} [-]	d_w [mm]
A	1	5	1.023	3.697	0.730	2.697	0.4476	1.020	1.70	0.7609
	2	5	0.897	2.792	0.642	1.792	0.2500	0.840	1.95	0.4885
	3	2	0.298	2.548	0.608	1.548	0.0685	0.267	2.05	0.1400
	4	3	0.455	2.121	0.528	1.121	0.0588	0.362	2.27	0.1334
	5	1	0.231	2.392	0.582	1.392	0.0446	0.200	2.05	0.0913
	6	1	0.196	3.280	0.695	2.280	0.0725	0.191	1.79	0.1299
B	1	20	1.467	2.814	0.645	1.814	0.4164	1.371	2.00	0.8328
	2	20	1.412	2.225	0.551	1.225	0.2174	1.167	2.21	0.4805
	3	20	0.283	2.638	-	-	0.0800	0.346	2.09	0.1686
	4	20	0.417	2.185	-	-	0.0728	0.490	2.30	0.1677
	5	20	0.222	2.449	-	-	0.0535	0.269	2.20	0.1177
	6	20	0.193	3.348	-	-	0.0809	0.237	1.88	0.1520
C	1	50	2.843	2.376	0.579	1.843	0.5405	2.469	2.13	1.1486
	2	50	3.609	1.921	0.480	0.921	0.2959	2.587	2.43	0.7176
	3	50	0.283	2.638	-	-	0.0109	3.975	-	-
	4	50	0.417	2.185	-	-	0.0110	3.975	-	-
	5	50	0.222	2.449	-	-	0.0750	3.975	-	-
	6	50	0.193	3.348	-	-	0.1010	3.975	-	-

Table 6.5: Estimates of the Kozeny-Carman coefficients using analytical interpretation of the particle-size distribution function and considering stone and gravel content for six sandy materials of (A) the original layered, (B) the refilled, and (C) the rubble material.

where h is pressure head, θ is volumetric water content, θ_s is the saturated water content, often assuming $\theta = 0.9n$ (0.1 is then the residual saturation of air, s_{ol}), and $F(d)$ is the cumulative particle size distribution function. The relation $d = \gamma R$ describes the relation between the grain diameter d and the equivalent pore radius R with $2R$ is equivalent to the effective hydraulic diameter. The packing index γ may vary between $\gamma = 4.83$ and $\gamma = 8.89$.

Pedotransfer Function Procedure

After curve-fitting of the estimated particle-size distribution in (6.9)–(6.11) the optimized parameters m, μ, and d_g are obtained. Using the estimated bulk density ρ_b, one can calculate an estimate of the pore-distribution index

$$\hat{\lambda} = a_1 \mu \rho_b^{a_2}. \tag{6.27}$$

The coefficients $a_1 = 0.0723$ and $a_2 = 3.8408$ are valid for values of ρ_b between $1.5 \leq \rho_b \leq 1.75$ and $\mu = m/(1-m)$ is calculated from (6.9) (see Table 6.5).

The packing index γ, is the estimated from

$$\gamma = b_1 + b_2 \hat{\lambda} + b_3 \hat{\lambda}^2, \tag{6.28}$$

where b_1, b_2, and $b_3 = 0.1589$. For the desorption curve the air-entry value h_{ae}, is calculated by

$$\frac{h_{ae}}{\gamma} = \frac{0.149}{d_g} \left[\left(\frac{\theta_{ae}}{\theta_s} \right)^{-\frac{1}{m}} - 1 \right]^{1-m} \frac{h_{ae}}{h(\theta_{ae})} \tag{6.29}$$

using the relation

$$\frac{h_{ae}}{h(\theta_{ae})} = \left[(1 + \hat{\lambda}) \left\{ 1 - \frac{h_{ae}}{h(\theta_{ae})} \left(1 - \frac{\theta_s}{\varepsilon} \right) \right\} \right]^{-\frac{1}{\hat{\lambda}}} \tag{6.30}$$

with

$$\theta_{ae} = \frac{\varepsilon}{1 + \hat{\lambda}} \tag{6.31}$$

and

$$\varepsilon = 1 - \frac{\rho_b}{\rho_f}, \tag{6.32}$$

where (6.30) is solved iteratively. The equation for the main desorption branch of the function is

$$\begin{array}{ll} \theta = \varepsilon \left[\frac{h_{ae}}{h} \right]^\lambda \left[1 - \frac{h_{ae}}{h} \left(1 - \frac{\theta_s}{\varepsilon} \right) \right], & h > h_{ae} \\ \theta = \theta_s, & h \leq h_{ae}. \end{array} \tag{6.33}$$

For calculating the desorption curve according to Haverkamp and Parlange, we used the program PTF of Tietje et al. [ea91] to obtain a table of discrete values of the pressure-saturation relation for the six different sandy soil types. By curve fitting of the estimated $\theta(h)$-relations (6.13)–(6.16), using the optimization program RETC (van Genuchten et al. [vGLY91]), we obtained the model parameters presented in Table 6.7.

Since the analysis of Haverkamp and Parlange is based on the Brooks and Corey model, the fit using the BC-model is better than using the VGM-model especially for coarse sands. For the medium and fine sands the fit of the BC-model is relatively bad. Results are summarized in Tables 6.7 and 6.8.

6.1.12 Discussion

The fit of the particle-size distribution function $F(d)$ to the descriptions given in the soil protocol results in estimated values which are not optimal for the sandy materials. Especially the portions of the fine grain sizes (silt and clay) mostly remain underestimated by the curve-fitting procedure. For the mostly sandy soils such an underestimation has a relatively large effect on the parameter estimation since the total content of fine material is already relatively low and thus highly sensitive. For the coarse sand the BC-model seems to be nearly ideally suited, however, not for the medium and fine sands where the coefficients $\lambda \to 0$ and $\alpha \to 0$ are approaching zero. For soils with differing bulk densities the curve shape will be different depending on the packing index γ in the model of Haverkamp and Parlange, which means that for each different bulk density a new estimate of the pedotransfer function is required. For the refilled and rubble material the parameters in Tables 6.7 and 6.8 can be regarded as estimates of the pressure-saturation relation as a function of the relative saturation $S_e = S_e(h)$. The absolute air and water contents are obtained by multiplication with each phase-filled porosities, n.

6.1. Pedotransfer Model

	soil material	ρ_b $[\frac{g}{cm^3}]$	T^* [-]	n^* [-]	n [-]	n' [-]	d_w $[10^{-3}m]$	K_w $[10^{-12}m^2]$
A	1	1.46	2.25	0.449	0.427	0.405	0.7609	655.87
	2	1.49	2.25	0.438	0.416	0.387	0.4885	227.06
	3	1.52	2.40	0.426	0.417	0.334	0.1400	10.28
	4	1.60	2.40	0.396	0.384	0.230	0.1334	3.00
	5	1.44	2.30	0.457	0.452	0.362	0.0913	6.57
	6	1.54	2.30	0.419	0.415	0.290	0.1299	6.00
B	1	1.46	2.25	0.449	0.359	0.341	0.8328	374.74
	2	1.49	2.25	0.438	0.350	0.326	0.4805	106.00
	3	1.52	2.40	0.426	0.341	0.273	-	4.39
	4	1.60	2.40	0.396	0.317	0.190	-	1.37
	5	1.44	2.30	0.457	0.366	0.292	-	2.58
	6	1.54	2.30	0.419	0.335	0.235	-	2.47
C	1	1.46	2.25	0.449	0.225	0.213	-	52.16
	2	1.49	2.25	0.438	0.219	0.204	-	18.57
	3	1.52	2.40	0.426	0.213	0.170	-	0.744
	4	1.60	2.40	0.396	0.198	0.119	-	0.245
	5	1.44	2.30	0.457	0.229	0.183	-	0.429
	6	1.54	2.30	0.419	0.210	0.147	-	0.429

Table 6.6: Estimates of bulk density ρ_b, tortuosity coefficient T^*, porosities n^*, n, n', effective hydraulic particle diameter d_w, and specific permeability K_w, according to Kozeny-Carman equation for six sandy materials of the originally layered, refilled, and rubble substrates for (A) the original layered, (B) the refilled, and (C) the rubble material.

	soil material	θ_s [-]	α $[cm^{-1}]$	n [-]	λ [-]
A	1	0.449	0.4976	3.551	-
	2	0.438	0.7860	1.703	-
	3	0.426	0.2789	1.596	-
	4	0.396	0.4047	1.514	-
	5	0.457	0.2501	1.422	-
	6	0.419	0.1729	1.912	-
A	1	0.449	0.8670	-	1.2303
	2	0.438	0.6892	-	1.4520

Table 6.7: Estimated parameters of the pressure-saturation function according to the VGM- $(n, m, m = 1 - 1/n)$ and BC-model (λ) using curve-fitting program RETC for loosely packing of originally layered material with $\theta_r = 0.0, l = 0.5$ and $K_s = 1$ for (A) the original layered sand.

	soil material	θ_s [-]	α [cm^{-1}]	n [-]
A	1	0.32	0.8167	12.951
	2	0.32	0.5173	2.675
	3	0.31	0.1960	2.260
	4	0.27	0.2709	2.006
	5	0.34	0.1722	1.714
	6	0.30	0.1173	3.191

Table 6.8: Estimated values of the pressure-saturation function according to the VGM-model $(n, m, m = 1 - 1/n)$ for medium and densely packed originally layered materials assuming θ_s=porosity* (using $\theta_r = 0.0, l = 0.5$ and $K_s = 1$) for (A) the original layered sand.

6.2 Spatial Variability and Uncertainties

In the soil venting procedure there are many sources of uncertainties

1. Air transmissivity in the soil,
2. Boundary conditions for the gas flow,
3. NAPLs distribution in the soil,
4. NAPLs diffusivity and transfer coefficients.

We consider the optimization in the case of the stochastic transmissivity. We can suppose that the transmissivity is a random variable with some spatial structure [Gor90]. A simpler model is to treat it as a second-order stationary spatial process. Here the expected value of the quadratic variation between two transmissivities value, separated by a distance ΔX does not depend on the location of the values but only on their distance. Under the stationarity assumption and considering the transmissivity variation to be log-normally distributed, the spatial correlation can be defined using a variety of functions. Here we adopt an exponential semi-variogram γ [Gor90],

$$\gamma(\Delta X) = \sigma_{\log T}^2 (1 - e^{-|\Delta X|/\lambda}) \tag{6.34}$$

where $\sigma_{\log T}^2$ is the variance of the logarithms of the transmissivity and λ is the correlation length. Using equation (6.34) one can generate maps of the transmissivity. The same can be done for the NAPL concentration as well as for the transfer coefficient.

6.3 Ordinary Kriging

First we describe the general procedure of *kriging*.

Data: We suppose that w is a stationary random field of order 2 (cf. Journel, Huijbregts [JH78], page 32) with expectation $E\ w(x) = m$. Further, we assume that measured values of w_j $(j = 1, \ldots, n)$ at observation points x_j $(j = 1, \ldots, n)$ are given. The covariance function $R(h) = \text{Cov}\ (w(x), w(x+h))$ is known and it is "positive definite", i.e., the covariance matrix

$$\begin{pmatrix} R(0) & R(|x_1 - x_2|) & R(|x_1 - x_3|) & \ldots & R(|x_1 - x_n|) \\ R(|x_2 - x_1|) & R(0) & R(|x_2 - x_3|) & \ldots & R(|x_2 - x_n|) \\ \ldots & \ldots & \ldots & \ldots & \ldots \\ R(|x_n - x_1|) & R(|x_n - x_2|) & R(|x_n - x_3|) & \ldots & R(0) \end{pmatrix}$$

is positive definite (see Karlin, Taylor [KT75] Th. 7.1).

Unknown: The first problem is to determine an estimate w^* for the value $w(x^*)$ at a given point x^*.

Proposition 6.1 (Kriging) *Among all unbiased estimators of the form*

$$w^* = \sum_{j=1}^{n} \lambda_j w_j \tag{6.35}$$

the one with the least variance

$$\text{Var}\ (w^* - w(x^*))$$

of the estimation error is given by solving the system

$$\begin{cases} \sum_{j=1}^{n} R(|x_k - x_j|)\lambda_j + \mu = R(|x_k - x^*|) & \forall k \\ \sum_{j=1}^{n} \lambda_j = 1 \end{cases} \tag{6.36}$$

for the weights λ_j and the Lagrange multiplier μ.

Proof: See Proposition 6.4. □

Proposition 6.2 *Let $w^* = \sum_{j=1}^{n} \lambda_j w_j$, where λ_j are given by (6.36). Then*

$$w^* = \sum_{j=1}^{n} R(|x^* - x_j|)\alpha_j + \beta,$$

where $\alpha_j\ \forall j, \beta$ are given by solving the system

$$\begin{cases} \sum_{j=1}^{n} R(|x_k - x_j|)\alpha_j + \beta = w_k & \forall k \\ \sum_{j=1}^{n} \alpha_j = 0. \end{cases} \tag{6.37}$$

Proof: Let us note that the algebraic systems (6.36) and (6.37) have the same matrix A. This is symmetric and positive definite. Let its inverse be

$$A^{-1} = (b_{i,j})_{i,j=1,\dots,n+1}.$$

Then we can write

$$\begin{cases} \lambda_j = b_{j,n+1} + \sum_{i=1}^{n} b_{j,i} R(|x_i - x^*|) \\ \mu = b_{n+1,n+1} + \sum_{i=1}^{n} b_{n+1,i} R(|x_i - x^*|), \end{cases}$$

and

$$\begin{cases} \alpha_j = \sum_{i=1}^{n} b_{j,i} w_i \\ \beta = \sum_{i=1}^{n} b_{n+1,i} w_i. \end{cases}$$

This implies

$$w^* = \sum_{j=1}^{n} \lambda_j w_j = \sum_{j=1}^{n} w_j \left[b_{j,n+1} + \sum_{i=1}^{n} b_{j,i} R(|x_i - x^*|) \right]$$

$$= \sum_{i=1}^{n} R(|x_i - x^*|) \sum_{j=1}^{n} b_{j,i} w_j + \sum_{j=1}^{n} b_{j,n+1} w_j$$

$$= \sum_{i=1}^{n} R(|x_i - x^*|) \alpha_i + \beta. \qquad \square$$

Remark 6.3 According to Proposition 6.2 one has to solve only one system of linear equations, namely (6.37). $\square$

Proposition 6.4 *If the observations have independent errors e_j with zero expectation and variances ε_j, then the system to be solved is*

$$\begin{cases} \sum_{j=1}^{n} R(|x_k - x_j|) \lambda_j + \varepsilon_k \lambda_k + \mu = R(|x_k - x^*|) \quad \forall k \\ \sum_{j=1}^{n} \lambda_j = 1 \end{cases} \qquad (6.38)$$

Proof: We have

$$\frac{1}{2} \text{Var}\,(w^* - w(x^*)) = \frac{1}{2} \text{E}\,\left(\sum_{j}^{n} \lambda_j w_j - w(x^*)\right)^2$$

$$= \frac{1}{2} \sum_{k,j=1}^{n} \lambda_k \lambda_j \text{Cov}\,(w_k, w_j) - \sum_{j=1}^{n} \lambda_j \text{Cov}\,(w_j, w(x^*)) + \frac{1}{2} \text{Var}\,(w(x^*))$$

6.3. Ordinary Kriging

Since we know
$$\text{Cov}\,(w_k, w_j) = R(|x_k - x_j|) + \delta_{k,j}\varepsilon_k,$$
we have to minimize the expression
$$\frac{1}{2}\sum_{k,j=1}^{n}\lambda_k\lambda_j(R(|x_k - x_j|) + \delta_{k,j}\varepsilon_k) - \sum_{j=1}^{n}\lambda_j R(|x_j - x^*|)$$

If λ is the vector with components λ_j, $\tilde{A}$ is the matrix with elements
$$R_{k,j} = R(|x_k - x_j|) + \delta_{k,j}\varepsilon_k,$$
and b is the vector with components $b_k = R(|x_j - x^*|)$, we can write this as
$$\frac{1}{2}\lambda^{tr}\tilde{A}\lambda - b^{tr}\lambda.$$

The gradient of this function with respect to λ is given by
$$\tilde{A}\lambda - b,$$
therefore, the optimal λ together with the Lagrange multiplier μ must satisfy the system
$$\tilde{A}\lambda + \mu = b. \qquad \square$$

Proposition 6.5 *Let* $w^* = \sum_{j=1}^{n}\lambda_j w_j$, *where* λ_j *are given by (6.38). Then*
$$w^* = \sum_{j=1}^{n} R(|x^* - x_j|)\alpha_j + \beta,$$
where α_j $(j = 1, \ldots, n)$ *and* β *are given by solving the system*
$$\begin{cases} \sum_{j=1}^{n} R(|x_k - x_j|)\alpha_j + \varepsilon_k\alpha_k + \beta &= w_k \quad \forall k \\ \sum_{j=1}^{n} \alpha_j &= 0 \end{cases} \qquad (6.39)$$

Proof: The proof can be done analogously as in the Proposition 6.2. $\qquad \square$

Remark 6.6 If the variance $\varepsilon(x)$ is independent of x, we may unify Proposition 6.1 and 6.4 by replacing the original function $R(h)$ by $R(h) + \varepsilon\delta_0(h)$, where $\delta_0(h)$ is the Dirac function with support at 0. $\qquad \square$

Estimating the Covariance

We assume that we have the same data as in Section 6.3. The main difficulty of kriging theory is to obtain the formula for covariance function R. In reality only measurements at some points are given. We describe one way how to obtain an estimator for R and demonstrate it on one example of Trichlorethen measurements in a polygonal domain in $\mathbb{R}^2$. See also Delhomme [Del78].

Let us define the semi-variogram function γ as follows

$$\gamma(h) = \frac{1}{2} \mathrm{E}\left([w(x+h) - w(x)]^2\right).$$

Under stationarity assumptions the semi-variogram function γ is equivalent to the covariance R (cf. Journel, Huijbregts [JH78], page 32)

$$\gamma(h) = R(0) - R(h) = \mathrm{Var}\,(w(x)) - R(h).$$

An estimator $2\gamma^*$ of 2γ is the arithmetic mean of the squared differences between two measurements $[w(x_j), w(x_j + h)]$ at any two points separated by the vector h

$$2\gamma^*(h) = \frac{1}{N(h)} \sum_{i=1}^{N(h)} [w(x_j) - w(x_j + h)]^2,$$

where $N(h)$ is the number of experimental pairs $[w(x_j), w(x_j + h)]$ separated by the vector h. If $N(h) = 1$ then the graph is shown in Figure 6.2 (a). When the experimental

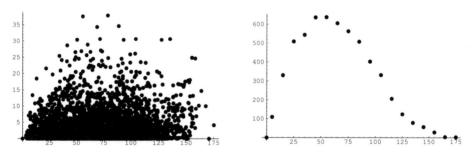

(a) Chaotic distribution – class length 0 m (b) Class counter – class length 10 m

Figure 6.2: Squared differences $[w(x_j) - w(x_j + h)]^2$ of measured data versus spatial distance h between the samples.

points are irregularly spaced in the domain, all pairs are regrouped into classes with some length (the choice of class-length is subjective). In our case, the maximal distance of two measurement points is $|h_{max}| = 175\ m$. We divide all pairs into classes with the class-length 10 m. The corresponding class-counter is shown in Figure 6.2 (b). We

6.4. Generation of a Random Field

see that the values of class-counter are between 0 and 650. Hence $2\gamma^*$ will be a good estimator of 2γ (from statistical point of view) for large values of class-counter. Let us consider classes with class-counter larger than 300, only. For each class we have computed $2\gamma^*$ and $R^*(h) = \text{Var}(w(x)) - \gamma^*$. The choice of the form of the function R^* is subjective, but R^* must be positive definite, i.e.,

$$\int_0^T \psi(t) \int_0^t R^*(t-s)\psi(s)\,ds\,dt \geq 0$$

for every $\psi \in C([0,\infty))$ and every $T > 0$ (see [GS80] pp. 208), moreover R^* is said to be strongly positive definite if there is an $\varepsilon > 0$ such that the function $t \mapsto R^*(t) - \varepsilon e^{-t}$ is positive definite. Strong positivity implies positivity. Using Laplace transform one can show that $R^* \in L^1(0,\infty)$ is strongly positive definite if and only if (see Nohel and Shea [NS76])

$$\text{Re } \widehat{R^*}(is) \geq \frac{\varepsilon}{1+s^2} \qquad \forall s \in \mathbb{R},$$

where $\widehat{R^*}$ is the Laplace transform of R^*. From the viewpoint of applications, it is useful to know some sign conditions for R^* which imply strong positiveness, e.g. (see Staffans [Sta76])

$$(-1)^j R^{*(j)}(t) \geq 0 \qquad \forall t \geq 0, \qquad j = 0, 1, 2; \ R^{*'} \neq 0.$$

For our purposes we have chosen the following form for R^*

$$R^*(h) = a \exp\left(bh + \frac{ch}{1+dh^2}\right).$$

Statistically relevant points of R^* (black color) are approximated by R^* using the least squares method. The whole situation is drawn in Figure 6.3. Statistically irrelevant points of R^* (i.e., class-counter is less than 300) are gray colored.

6.4 Generation of a Random Field

Now we are going to describe how Monte Carlo simulations can be generated.

Data: We assume that we have the same data as in Section 6.3.

Unknown: We consider the problem of determining Monte Carlo realizations $w(x)$ of the field w that satisfy two conditions:

1. $w(x)$ has the prescribed covariance structure given by the function R,

2. it satisfies $w(x_j) = w_j$ for all j.

Proposition 6.7 *Let $w^*(x)$ be the kriging function corresponding to the data (x_j, w_j). Let further $v(x)$ be any realization having the covariance structure R, and $v^*(x)$ the kriging function corresponding to the data (x_j, v_j) with $v_j = v(x_j)$. Then the function $\tilde{w}(x) = w^*(x) + v(x) - v^*(x)$ has the desired properties.*

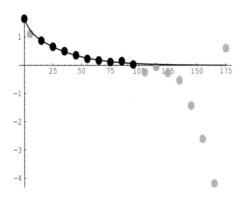

Figure 6.3: Estimator of a covariance. Fitted points are black and irrelevant points are gray colored.

Proof: (a) We have $v(x_j) = v^*(x_j)$ and therefore $w(x_j) = w^*(x_j) = w_j$.

(b) We obtain $\mathrm{E}\,\tilde{w}(x) = \mathrm{E}\,w^*(x)$.

(c) If $v = w$ then $\tilde{w} = w$, thus $\tilde{w}$ has the same covariance structure as w. Let us suppose that v is an independent realization of w. Then we can write

$$\mathrm{Cov}\,(w(x) - w^*(x), v^*(y)) = \mathrm{Cov}\,(v(x) - v^*(x), w^*(y)) = 0$$

for all x, y. According to (6.36) we have

$$\mathrm{Cov}\,(w(x) - w^*(x), w^*(y))$$
$$= \sum_{j=1}^{n} \lambda_j(y) \left[\mathrm{Cov}\,(w(x), w(x_j)) - \sum_{i=1}^{n} \lambda_i(x) \mathrm{Cov}\,(w(x_i), w(x_j)) \right]$$
$$= 0.$$

If w, v have the same covariance function then the kriging estimates $w^*(x), v^*(x)$ at point x will be computed with the same weights $\lambda_i(x), (i = 1, \ldots, n)$ (cf. (6.36)). Therefore, we get

$$\mathrm{Cov}\,(w^*(x), w(y)) = \sum_{i=1}^{n} \lambda_i(x) \mathrm{Cov}\,(w(x_i), w(y))$$
$$= \sum_{i=1}^{n} \lambda_i(x) \mathrm{Cov}\,(v(x_i), v(y))$$
$$= \mathrm{Cov}\,(v^*(x), v(y)).$$

6.4. Generation of a Random Field

Finally, we can write

$$\begin{aligned}
&\mathrm{Cov}\,(\tilde{w}(x), \tilde{w}(y)) \\
&= \mathrm{Cov}\,(w^*(x), w^*(y)) + \mathrm{Cov}\,(w^*(x), v(y) - v^*(y)) + \mathrm{Cov}\,(v(x) - v^*(x), w^*(y)) \\
&\quad + \mathrm{Cov}\,(v(x), v(y)) - \mathrm{Cov}\,(v^*(x), v(y)) - \mathrm{Cov}\,(v(x) - v^*(x), v^*(y)) \\
&= \mathrm{Cov}\,(w^*(x), w^*(y)) + \mathrm{Cov}\,(w(x), w(y)) - \mathrm{Cov}\,(w^*(x), w(y)) \\
&= \mathrm{Cov}\,(w(x), w(y)).
\end{aligned}$$
$\square$

There exist some multidimensional methods for generating random fields, but they are too expensive for computer simulations. The basic concept of the Turning Band Method (TBM) is to transform a multidimensional simulation into the sum of equivalent 1-D simulations. It preserves the statistics of the true field.

Definition 6.8 (Turning band method) *Let $\boldsymbol{\zeta}_j$ for $j = 1, \ldots, m$ denote uniformly distributed unit vectors on the unit circle S^1 in $\mathbb{R}^2$ or on the unit sphere S^2 in $\mathbb{R}^3$, resp. We assume that w_j is a family of mutually independent one-dimensional weakly stationary random fields with a zero mean and with the same covariance function R_1. Then a TBM field w in $\mathbf{R}^n$ is defined by the formula*

$$w(\mathbf{x}) = \frac{1}{\sqrt{m}} \sum_{j=1}^{m} w_j(\mathbf{x} \cdot \boldsymbol{\zeta}_j). \tag{6.40}$$

Proposition 6.9 *If w is generated from the family w_j having the covariance function R_1, then w is weakly stationary, has zero mean and in the limit $m \to \infty$ the n-dimensional covariance function*

$$R(\mathbf{x}) = \frac{1}{|S^{n-1}|} \int_{S^{n-1}} R_1(\mathbf{x} \cdot \boldsymbol{\zeta}) \, do(\boldsymbol{\zeta}). \tag{6.41}$$

Proof: Firstly we have

$$\mathrm{E}\,[w(\mathbf{x})] = \mathrm{E}\left[\frac{1}{\sqrt{m}} \sum_{j=1}^{m} w_j(\mathbf{x} \cdot \boldsymbol{\zeta}_j).\right]$$

Since the w_j are mutually independent, we obtain immediately

$$\mathrm{E}\,[w(\mathbf{x})] = \frac{1}{\sqrt{m}} \sum_{j=1}^{m} \mathrm{E}\,[w_j(\mathbf{x} \cdot \boldsymbol{\zeta}_j)],$$

and thus

$$\mathrm{E}\,[w(\mathbf{x})] = 0.$$

Secondly we have

$$\mathrm{Cov}\,(w(\mathbf{x}_1), w(\mathbf{x}_2)) = \mathrm{E}\,[w(\mathbf{x}_1)w(\mathbf{x}_2)] = \frac{1}{m}\mathrm{E}\left[\sum_{j=1}^{m} w_j(\mathbf{x}_1 \cdot \boldsymbol{\zeta}_j) \sum_{k=1}^{m} w_k(\mathbf{x}_2 \cdot \boldsymbol{\zeta}_k)\right].$$

We get
$$\text{Cov}\,(w(\mathbf{x}_1), w(\mathbf{x}_2)) = \frac{1}{m}\sum_{j=1}^{m}\sum_{k=1}^{m} \text{E}\,[w_j(\mathbf{x}_1 \cdot \boldsymbol{\zeta}_j)w_k(\mathbf{x}_2 \cdot \boldsymbol{\zeta}_k)].$$

Due to the mutual independence we obtain
$$\text{Cov}\,(w(\mathbf{x}_1), w(\mathbf{x}_2)) = \frac{1}{m}\sum_{j=1}^{m} \text{E}\,[w_j(\mathbf{x}_1 \cdot \boldsymbol{\zeta}_j)w_j(\mathbf{x}_2 \cdot \boldsymbol{\zeta}_j)] = \frac{1}{m}\sum_{j=1}^{m} R_1((\mathbf{x}_1 - \mathbf{x}_2) \cdot \boldsymbol{\zeta}_j).$$

This gives
$$R(\mathbf{x}) = \frac{1}{m}\sum_{j=1}^{m} R_1(\mathbf{x} \cdot \boldsymbol{\zeta}_j),$$

which tends to
$$R(\mathbf{x}) = \frac{1}{|S^{n-1}|}\int_{S^{n-1}} R_1(\mathbf{x} \cdot \boldsymbol{\zeta})\, do(\boldsymbol{\zeta})$$

for $m \to \infty$, since the directions $\boldsymbol{\zeta}_j$ are assumed to be uniformly distributed. $\square$

Proposition 6.10 *The covariance function R from formula (6.41) is isotropic, i.e., $R(\mathbf{x})$ is a function of $r = |\mathbf{x}|$. For $R(r)$ instead of $R(\mathbf{x})$ one has*

$$R(r) = \frac{1}{2\pi}\int_0^{2\pi} R_1(r\cos\psi)\, d\psi \quad \text{if } n = 2 \tag{6.42}$$

and

$$R(r) = \frac{1}{r}\int_0^r R_1(\tilde{r})\, d\tilde{r} \quad \text{if } n = 3. \tag{6.43}$$

Proof: For $n = 2$ formula (6.41) reads
$$R(r) = \frac{1}{2\pi}\int_0^{2\pi} R_1(r\cos\psi)\, d\psi = \frac{1}{\pi}\int_0^{\pi} R_1(r\cos\psi)\, d\psi$$

because of the symmetry of the function R_1. For $n = 3$ formula (6.41) reads
$$R(r) = \frac{1}{4\pi}\int_0^{2\pi}\int_0^{\pi} R_1(r\cos\vartheta)\sin\vartheta\, d\vartheta d\varphi.$$

Now we use the substitution $r\cos\vartheta = \tilde{r}$ and get the result. $\square$

Since $R(\mathbf{x})$ is a function of $r = |\mathbf{x}|$, its Fourier transform $\hat{R}(\boldsymbol{\xi})$ is a function of $\rho = |\boldsymbol{\xi}|$; therefore we write $\hat{R}(\rho)$ instead of $\hat{R}(\boldsymbol{\xi})$.

Notations: We use the Fourier transform formulas
$$\hat{R}(\boldsymbol{\xi}) = \int_{\mathbf{R}^n} R(\mathbf{x})e^{-2\pi i \boldsymbol{\xi}\cdot\mathbf{x}}\, d\mathbf{x}$$

and
$$R(\mathbf{x}) = \int_{\mathbf{R}^n} \hat{R}(\boldsymbol{\xi})e^{2\pi i \boldsymbol{\xi}\cdot\mathbf{x}}\, d\boldsymbol{\xi}.$$

6.4. Generation of a Random Field

For $n = 2$ we have

$$\hat{R}(\rho) = 2\pi \int_0^\infty R(r) r J_0(2\pi r \rho) \, dr$$

and

$$R(r) = 2\pi \int_0^\infty \hat{R}(\rho) \rho J_0(2\pi r \rho) \, d\rho$$

(see Abramowitz, Stegun [AS64] formula 9.1.18).

Proposition 6.11 *Let $n = 2$, let R and R_1 be related by formula (6.42), then one has for the one-dimensional Fourier transform $\widehat{R_1}$ of R_1*

$$\widehat{R_1}(\rho) = \pi \rho \hat{R}(\rho), \quad \rho \geq 0.$$

Proof: Without loss of generality we can assume that R_1 and $\widehat{R_1}$ are defined on the whole real line as even functions. We substitute the Fourier representation

$$R_1(s) = 2 \int_0^\infty \widehat{R_1}(\rho) \cos(2\pi s \rho) \, d\rho$$

into formula (6.42) and get

$$R(r) = \frac{1}{\pi} \int_0^{2\pi} \int_0^\infty \widehat{R_1}(\rho) \cos(2\pi r \rho \cos \psi) \, d\rho d\psi = \frac{1}{\pi} \int_0^\infty \widehat{R_1}(\rho) \int_0^{2\pi} \cos(2\pi r \rho \cos \psi) \, d\psi d\rho.$$

On the other hand, we have the Fourier representation

$$R(r) = \int_0^\infty \rho \hat{R}(\rho) \int_0^{2\pi} \cos(2\pi r \rho \cos \varphi) \, d\varphi d\rho.$$

Since the Fourier transform is unique, we get the result. $\square$

Proposition 6.12 *Let $n = 3$, let R and R_1 be related by formula (6.43), then one has*

$$R_1(r) = \frac{d}{dr}(rR(r))$$

Proof: trivial $\square$

After obtaining the covariance function of the unidimensional process, we can easily generate the process along the turning bands lines using any spectral method. We present two simple ways based on Fourier series or Fourier transform for generating 1-D random fields with a given covariance function. For some details, modifications or another approaches we refer the reader to Shinozuka and Jan [SJ72], Mantoglou and Wilson [MW82], Braud and Obled [BO91].

Proposition 6.13 *Let $A_0, \ldots, A_k$ and $B_0, \ldots, B_k$ be uncorrelated random variables having zero means. We assume that A_i, B_i $(i = 0, \ldots, k)$ have a common variance σ_i. Then*

$$w(x) = \sum_{i=0}^k \{A_i \cos(ipx) + B_i \sin(ipx)\}, \quad p, x \in \mathbb{R}, \ p > 0$$

has the zero mean and its covariance function is given by

$$\mathrm{Cov}\,(w(x), w(x+h)) = \sum_{i=0}^{k} \sigma_i \cos(iph).$$

Proof: We can write

$$\begin{aligned}
\mathrm{E}\,[w(x)] &= \mathrm{E}\left[\sum_{i=0}^{k} \{A_i \cos(ipx) + B_i \sin(ipx)\}\right] \\
&= \sum_{i=0}^{k} \{\mathrm{E}\,[A_i] \cos(ipx) + \mathrm{E}\,[B_i] \sin(ipx)\} = 0.
\end{aligned}$$

Since $A_0, \ldots, A_k$ and $B_0, \ldots, B_k$ are uncorrelated with zero means we have

$$\mathrm{E}\,[A_i B_j] = \mathrm{E}\,[A_i A_j] = \mathrm{E}\,[B_i B_j] = 0 \qquad \text{for } i \neq j.$$

For the covariance we obtain

$$\begin{aligned}
\mathrm{Cov}\,(w(x), w(x+h)) &= \mathrm{E}\,[w(x)w(x+h)] \\
&= \mathrm{E}\left[\left(\sum_{i=0}^{k} \{A_i \cos(ipx) + B_i \sin(ipx)\}\right)\left(\sum_{i=0}^{k} \{A_i \cos[ip(x+h)] + B_i \sin[ip(x+h)]\}\right)\right] \\
&= \sum_{i=0}^{k} \{\mathrm{E}\,[A_i^2] \cos(ipx)\cos(ip(x+h)) + \mathrm{E}\,[B_i^2] \sin(ipx)\sin(ip(x+h))\} \\
&= \sum_{i=0}^{k} \sigma_i \cos(iph).
\end{aligned}$$
$\square$

Example 6.14 Let $n = 2$ and a, b be given positive real numbers. Let us consider a simple covariance function

$$R(r) = a \exp(-br) \qquad r \geq 0.$$

We want to generate a 2-D random field with a given covariance function R such that it will have prescribed values at the measurement points.

Applying Gradshteyn and Ryzhik [GR80], p. 712, formula 6.623.2 we get

$$\hat{R}(\rho) = \frac{2\pi ab}{[b^2 + (2\pi\rho)^2]^{3/2}}.$$

Using Proposition 6.11 we have

$$\widehat{R_1}(\rho) = \frac{2\pi^2 ab\rho}{[b^2 + (2\pi\rho)^2]^{3/2}}.$$

6.4. Generation of a Random Field

Further, according to Gradshteyn and Ryzhik [GR80], p. 429, formula 3.773.4 we can write

$$R_1(r) = 2\int_0^\infty \widehat{R_1}(\rho)\cos(2\pi r\rho)\,d\rho = -\frac{\pi abr}{2}I_0(br) + a\,{}_1F_2\left(1;\frac{1}{2},\frac{1}{2};\frac{b^2r^2}{4}\right),$$

where $I_0(x)$ denotes the Bessel function and ${}_1F_2(a;b_1,b_2;x)$ is the generalized hypergeometric series (cf. Gradshteyn and Ryzhik [GR80], p. 1045, formula 9.14.1). The behavior of the function $R_1(r)$ for a given parameter set $a = 1.65, b = 0.03$ is shown in Figure 6.4 .

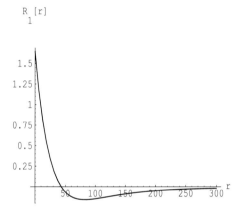

Figure 6.4: Covariance function $R_1(r)$ for 1-D stationary random field.

The covariance function R_1 is supposed to be insignificant outside the interval $(-L, L)$. Now, we expand R_1 into the Fourier series as

$$R_1(r) \approx \sum_{i=0}^{k} R_{1,i}\cos(ir\frac{\pi}{L}).$$

Let $\{A_i^j, B_i^j\}_{i=0,\ldots,k}^{j=1,\ldots,m}$ be uncorrelated random variables with zero means. Assume that A_i^j, B_i^j have a common variance $R_{1,i}$ for all $j = 1,\ldots,m$. Then according to Proposition 6.13

$$w_j(h) = \sum_{i=0}^{k}\left[A_i^j\cos(ih\frac{\pi}{L}) + B_i^j\sin(ih\frac{\pi}{L})\right] \qquad h \in \mathbb{R}$$

has a covariance function R_1 for all $j = 1,\ldots,m$. The TBM field w in $\mathbb{R}^2$

$$w(\mathbf{x}) = \frac{1}{\sqrt{m}}\sum_{j=1}^{m} w_j(\mathbf{x}\cdot\boldsymbol{\zeta}_j)$$

has a covariance function R. Finally, applying Proposition 6.7 we obtain a 2-D random field with a prescribed covariance function R keeping the measurement values at the given measurement points. □

The disadvantage of the just presented method (based on Fourier series) is, that one needs to use Fourier and inverse Fourier transforms. This can be done for simple covariance formulas, only. To avoid the troubles with the inverse Fourier transform we present an another spectral method for generating a random field with a prescribed covariance function.

Proposition 6.15 Let $A_0, \ldots, A_k$ be independent random angles uniformly distributed over $(0, 2\pi)$. Then

$$w(x) = \sum_{i=0}^{k} a_i \cos(b_i x + A_i), \qquad a_i, b_i, x \in \mathbb{R},$$

has the zero mean and its covariance function is given by

$$\mathrm{Cov}\,(w(x), w(x+h)) = \frac{1}{2} \sum_{i=0}^{k} a_i^2 \cos(b_i h).$$

Proof: One can write

$$E[w(x)] = \sum_{i=0}^{k} a_i E\left[\cos(b_i x + A_i)\right] = 0.$$

For the covariance function we have

$$\mathrm{Cov}\,(w(x), w(x+h)) = E\left[w(x) w(x+h)\right]$$

$$= E\left[\sum_{i=0}^{k} a_i \cos(b_i x + A_i) \sum_{j=0}^{k} a_j \cos(b_j (x+h) + A_j)\right]$$

$$= \sum_{i=0}^{k} a_i^2 E\left[\cos(b_i x + A_i)\cos(b_i (x+h) + A_i)\right]$$

$$= \frac{1}{2} \sum_{i=0}^{k} a_i^2 E\left[\cos(b_i x + b_i (x+h) + A_i) + \cos(b_i h)\right] = \frac{1}{2} \sum_{i=0}^{k} a_i^2 \cos(b_i h). \quad \square$$

Example 6.16 Let $n = 2$ and the covariance function and its Fourier transform be given by the Table 6.9. Using Proposition 6.11 we have

$$\widehat{R_1}(\rho) = \pi \rho \hat{R}(\rho), \quad \rho \geq 0.$$

We suppose that $\widehat{R_1}$ is insignificant outside the interval $(-L, L)$. Let $\{A_i^j\}_{i=0,\ldots,k}^{j=1,\ldots,m}$ be uncorrelated random variables uniformly distributed over $(0, 2\pi)$. Let us define

$$w_j(x) = 2 \sum_{i=0}^{k} \sqrt{\widehat{R_1}\left[\frac{L}{k+1}\left(i + \frac{1}{2}\right)\right] \frac{L}{k+1}} \cos\left[\frac{2\pi L x}{k+1}\left(i + \frac{1}{2}\right) + A_i^j\right],$$

$R(r)$	$\widehat{R}(\rho)$	Reference
$\dfrac{a}{r}$	$\dfrac{a}{\rho}$	[GR80], p. 665, formula 6.511.1
$\dfrac{a}{r^2+b^2}$	$2\pi a K_0(2\pi\rho b)$	[GR80], p. 678, formula 6.532.4
ae^{-br}	$\dfrac{2\pi ab}{[b^2+(2\pi\rho)^2]^{3/2}}$	[GR80], p. 712, formula 6.623.2
$\dfrac{a}{r}e^{-br}$	$\dfrac{2\pi a}{\sqrt{b^2+(2\pi\rho)^2}}$	[GR80], p. 707, formula 6.611.1
$arK_1(br)$	$\dfrac{4\pi ab}{[b^2+(2\pi\rho)^2]^2}$	[GR80], p. 675, formula 6.525.2

Table 6.9: 2-D Covariance Functions and their Fourier Transforms.

for $j = 1, \ldots, m$. According to Proposition 6.15 we have

$$\operatorname{Cov}(w_j(x), w_j(x+h))$$
$$= 2 \sum_{i=0}^{k} \widehat{R}_1 \left[\frac{L}{k+1}\left(i+\frac{1}{2}\right) \right] \cos\left[\frac{2\pi L h}{k+1}\left(i+\frac{1}{2}\right) \right] \frac{L}{k+1}$$
$$\xrightarrow{k\to\infty} 2 \int_0^L \widehat{R}_1(x)\cos(2\pi x h)\,dx \approx 2\int_0^\infty \widehat{R}_1(x)\cos(2\pi x h)\,dx = R_1(h).$$

The TBM field w in $\mathbb{R}^2$

$$w(\mathbf{x}) = \frac{1}{\sqrt{m}} \sum_{j=1}^{m} w_j(\mathbf{x} \cdot \boldsymbol{\zeta}_j)$$

has a covariance function R. Now using Proposition 6.7 we obtain a 2-D random field with a prescribed covariance function R and with the fixed values at the given points. □

Remark 6.17 Error analysis for TBM was discussed in Mantoglou and Wilson [MW82]. □

6.5 Parameter Identification of Kertess

Lack of data is an obvious problem by soil venting. The transmissivity of the soil matrix, location, nature and the quantity of NAPLs always remain unknowns. This constitute one handicap for modeling the NAPL transport in porous media. To overcome this difficulty many methods were developed to build this data using some measurements. These methods are called calibration, parameter identification or inverse problem and they are often applied to ill posed problems. Due to the available data, we do our parameter identification in the two following steps.

Layer	$\theta_r[-]$	$\theta_s[-]$	K_s [$10^{-12}m^2$]	$\alpha[cm^{-1}]$	$l[-]$	$n[-]$
Coarse Sand	0	0.449	655.87	0.4976	0.5	5.551
Middle Sand	0	0.426	10.28	0.2789	0.5	1.596
Fine Sand 1	0	0.457	6.57	0.2501	0.5	1.422
Fine Sand 2	0	0.419	6.00	0.1729	0.5	1.912

Table 6.10: Van Genuchten parameters for water ($m = 1 - \frac{1}{n}$).

Vertical Averaging of the Transmissivity

The goal of this section is to describe the way how to determine the air transmissivity. For the standard two-phase (air and water) flow in an unsaturated porous media, the soil consists of the gas and water phase and the porous matrix. So, for saturations we have

$$\theta_{gas} + \theta_{water} + \theta_{matrix} = 1.$$

The relation between saturation θ_{water} and the pressure head h is described by the van Genuchten formula (6.13) (cf. van Genuchten [van80]) the parameters of which are given in Table 6.10 for different soil-layers. The saturation of water (air) as a function of the pressure head h is shown in Figure 6.5. The air permeability k_{gas} can be computed as

$$k_{gas} = k_{water} \frac{\mu_{water}}{\mu_{gas}},$$

where μ_{water} and μ_{gas} are the dynamic viscosities of water and air, respectively. In this way we can compute the air permeability $k_{gas}(x, y, z)$ as a function of the position.

When the 3-D domain has a vertical thickness small compared to the horizontal lengths and the geological structure essentially consists of horizontal layers, then the air flow field in the porous matrix is also essentially horizontal, i.e. two-dimensional. Under these assumptions the air transmissivity can be computed by

$$T_{gas}(x, y) = \int_{z_1}^{z_2} k_{gas}(x, y, z) dz$$

where z_1 denotes the water table and z_2 the soil surface.

This method can be applied at the measurements points, only. The number of these is limited and non-uniformly distributed in the domain. Thus we have used the method of *kriging* in order to obtain the air transmissivity at an arbitrary point. The transmissivity values at the measurements points lie between $2.7 - 4.7 \cdot 10^{-11} \frac{m^2 s^3}{kg}$, i.e. the domain is relatively homogeneous. For the *kriging procedure* we have taken $\log T$ values instead of T values.

In Kertess, we have computed, the whole situation is complicated by the presence of buildings with basements of $2\ m$ depth. The water table is $5\ m$ under the soil surface. The location of the measurement points is shown in Figure 6.6. From this reason we have divided the domain horizontally into two zones

6.5. Parameter Identification of Kertess

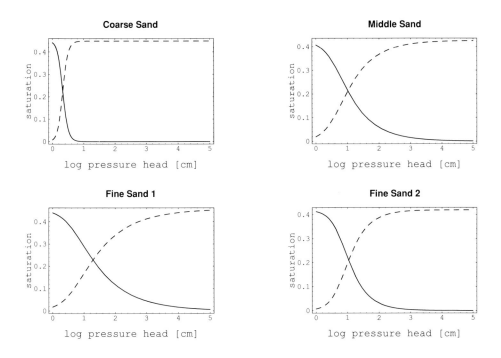

Figure 6.5: Saturation $\theta(h)$ for different layers (air corresponds to the solid and water to the dashed line).

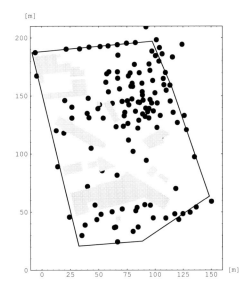

Figure 6.6: Kertess: Spatial distribution of measurement points.

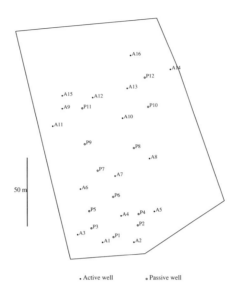

Figure 6.7: Kertess: Positions and numeration of wells used for calibration of flow parameters.

1. from the water table up to the basements of the buildings
2. the rest up to the soil surface

For each zone we have applied the kriging procedure separately neglecting the basements at the upper zone. After this, we have replaced the transmissivity at the buildings in the upper zone by 0. Finally we have added both particular transmissivities in order to obtain the total air transmissivity of the 2-D domain. Because of the vanishing transmissivity at the buildings in the upper zone we can observe sharp contours of all buildings in Figure 6.9.

6.5.1 Calibration of the Transmissivity T, the Leakage L and the Boundary Leakage l

1. **Data:** To calibrate the transmissivity and the leakage term, we make use of the air flow equation (4.4) and of the following measured data

    (a) The discharges of the active and passive wells $\widetilde{u}_i, i = 1, \ldots, N$ as shown in Figure 6.8 (a) and (b), respectively

    (b) The pressure at the passive wells $\widetilde{y}_i (i = n+1, \ldots, N)$ is equal to the atmospheric pressure

    where the positions of the wells in the domain are shown in Figure 6.7.

6.5. Parameter Identification of Kertess

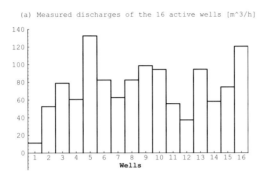

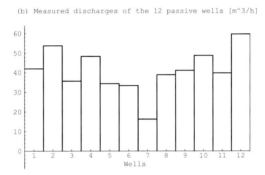

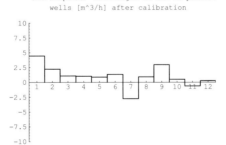

Figure 6.8: Kertess: Discharges for active and passive wells.

2. Parameters:

(a) **The transmissivity** T. Using kriging (see Section 6.3), we build a transmissivity field T_k from some measurements. Then we have to identify the piecewise constant function C_T defined as

$$C_T(\mathbf{x}) = \begin{cases} c_{T,1} & \text{if } \mathbf{x} \in \text{ a building with a deep (2 m) basement} \\ c_{T,2} & \text{if } \mathbf{x} \in \text{ a building with a middle deep} \\ & \text{(0.3 m) basement} \\ c_{T,3} & \text{otherwise} \end{cases} \quad (6.44)$$

with the constraint that there exist two strictly positive constants $\underline{c}$ and $\bar{c}$ such that

$$\underline{c} \le c_{T,1} \le c_{T,2} \le c_{T,3} \le \bar{c} \text{ for } i = 1,2,3$$

and we put

$$T(x,y) = C_T(x,y) * T_k(x,y)$$

(b) **The leakage term** L. We consider the leakage term as a piecewise constant function defined by

$$L(\mathbf{x}) = \begin{cases} L_1 & \text{if } \mathbf{x} \in \text{ a building} \\ L_2 & \text{if } \mathbf{x} \in \text{ a covered surface but there is no building} \\ L_3 & \text{if } \mathbf{x} \in \text{ a not covered surface} \end{cases} \quad (6.45)$$

subject to the constraint

$$\underline{L} \le L_1 \le L_2 \le L_3 \le \bar{L}$$

where $\underline{L}$ and $\bar{L}$ are given and strictly positive constants. This constraint simply says that the air penetrated through the surface into the soil is more important in the regions which are not covered and less important in the covered regions and under the buildings.

(c) **The leakage term l appearing in the transmission condition on the boundary in problem (4.4).** At the boundary of the domain, there is an impervious wall in the vertical direction which is $26 \, m$ deep but does not start exactly from the surface. In the north and the west of the domain, the wall starts after $1 \, m$ under the surface and in the rest after $0.3 \, m$. We take l as a piecewise constant function defined by

$$l(\mathbf{x}) = \begin{cases} l_1 & \text{if } \mathbf{x} \in \text{ the north or the west of } \Gamma \\ l_2 & \text{otherwise} \end{cases} \quad (6.46)$$

subject to the constraint

$$\underline{l} \le l_2 \le l_1 \le \bar{l}$$

where $\underline{l}$ and $\bar{l}$ are given and strictly positive constants. This constraint indicates the fact that the air penetrated through the boundary into the soil is more important in the less insulated parts, namely the north and the west of Γ.

6.5. Parameter Identification of Kertess

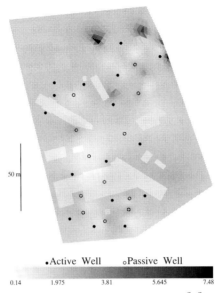

Figure 6.9: Kertess: Transmissivity T $[10^{-10}\frac{m^2 s^3}{kg}]$ after calibration.

All buildings, covered and not covered zones in the domain are shown in Figure 6.10. This formulation of the transmissivity and the leakage term leads to 8 constants to be found. We denote them by p_f;

$$p_f = \{c_{T,i}, i = 1, 2, 3, \; L_i, i = 1, 2, 3 \text{ and } l_i, i = 1, 2\}$$

and estimate them using the air flow model and the following objective functional.

3. **Objective functional:** The air flow parameters p_f are estimated using the generalized least squares model fit criterion. We minimize the functional J_f given by

$$J_f(p_f) \;=\; \sum_{i=n+1}^{N} W_{u_{ii}} (\hat{u}_i(p_f) - \widetilde{u}_i)^2 \tag{6.47}$$

with the notations : ($\sim$) means measured and ($\wedge$) means computed. As mentioned before, n is the number of active wells and $N - n$ is the number of passive wells. Here, the objective functional is expressed as the error in the discharges of the passive wells. The pressure of the passive wells and the discharges of the active wells are used as input data in the air flow problem.

6.5.2 Second Calibration of the Transmissivity T

The first calibration above of the transmissivity T, the leakage term L and the boundary leakage l is known to be zone approach. The domain and its boundary are divided into several regions and each of the three parameters T, L and l is corrected by one factor

Figure 6.10: Kertess: Leakage term L $[10^{-10}\frac{1}{m^2}]$ after calibration.

in each region. This global way to identify the parameters leads to poor approximation of the discharges at the passive wells. Therefore, we calibrate the transmissivity one second time using the same objective function (6.47) as criterion and the transmissivity at the passive well as unknown. The transmissivity in the whole domain is built using kriging. The difference between computed and measured discharges of the passive wells after the calibration is presented in Figure 6.8 (c).

6.5.3 Results of the Calibration of the Air Flow Parameters

Using discharges and the boundary conditions given by Figure 6.8 and the calibrated parameters, we get the pressure and the flow field represented by Figures 6.11 and 6.12.

6.5.4 Calibration of the Volatilization Coefficient V

Estimating the Concentration of the Pollutant and Volatilization

There are many kinds of NAPLs in the domain. The most important (quantitative and qualitative according to the measurements) is the Trichlorethen. The measurement values of the pollution of the soil matrix lie between 47 $\frac{\mu g}{m^3}$ − 2.5 · 10^7 $\frac{\mu g}{m^3}$. From these we have obtained an estimator of the covariance. For the *kriging procedure* we have taken *log* values of the measurements and the result of *kriging* is shown in Figure 6.15. The Figure 6.15 shows that there exist a few *hot spots* which have to be cleaned intensively. For the computation of extraction we suppose that the concentration of Trichlorethen

6.5. Parameter Identification of Kertess

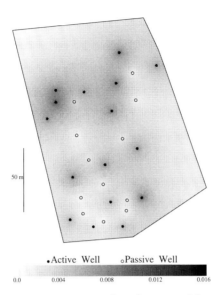

Figure 6.11: Kertess: Pressure y $[atm]$ using calibrated parameters.

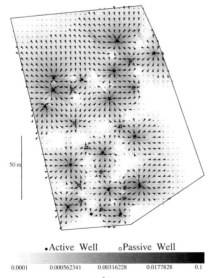

Figure 6.12: Kertess: Flux $\mathbf{q}$ $[\frac{kgm}{s}]$ using calibrated parameters.

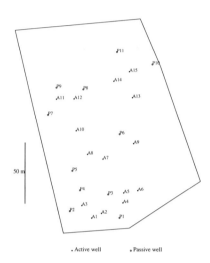

Figure 6.13: Kertess: Positions and numeration of the wells used for calibration of the volatilization coefficient and for the optimization.

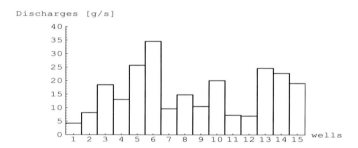

Figure 6.14: Kertess: Original discharges $[\frac{g}{s}]$ of active wells with the total discharge 240 $\frac{g}{s}$.

in the air-phase depends linearly on the concentration at the soil matrix. Hence we can write for the volatilization coefficient

$$V(\mathbf{x}) = V^* \frac{M(\mathbf{x}) - M_{\min}}{M_{\max} - M_{\min}},$$

where $V^* = 4 \cdot 10^{-7}\ \frac{kg}{m^2 s}$ and $M(\mathbf{x}), M_{\min}, M_{\max}$ denote the concentration of Trichlorethen at the soil matrix after the kriging procedure.

We suppose that the flow parameters T, L and l are now given and fixed. In the transport equation (3.6), we neglect the diffusivity and we assume that the saturation value for the relative pollutant density η_S is given. The pollutant density in the total extracted air is measured and available. Three NAPL components in our case are impor-

6.5. Parameter Identification of Kertess

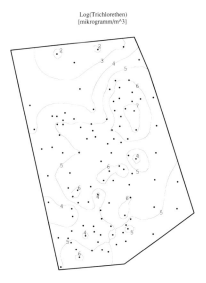

Figure 6.15: Kertess: Result of kriging – isolines.

NAPL component	Molar mass	Vapor density [kpa] at 20 °C	C_{sat} : Saturated density [kg/m^3]	C : Measurement[g/m^3]	V : Relative volatil.
cis-1,2-Di-chlorethen	96.94	24	0.9545	160.1	1.2
Tetra-chlorethen	165.8	1.9	0.129	2241.33	3.8
Tri-chlorethen	131.4	5.78	0.312	1035.28	11

Table 6.11: Chemical and physical properties of some NAPL components.

tant: cis-1,2-Dichlorethen, Tetrachlorethen and Trichlorethen. Some of their chemical and physical properties are given in Table 6.11.

The saturated density is obtained from the vapor density using the ideal gas law. We have
$$C_{sat} = \frac{PM}{RT}$$
where $T = 293.15\,K$ is the temperature, $R = 8.3146\,J$ is the ideal gas constant, P is the vapor density given in Table 6.11 and M is the molar mass of the component. The density η_i, $i = 1, 2, 3$ of each of the three components satisfies the equation
$$\nabla \cdot (\mathbf{q}\eta_i) = AV_i(\eta_{i\,s} - \eta_i)$$

The calibration of V is done as follows.

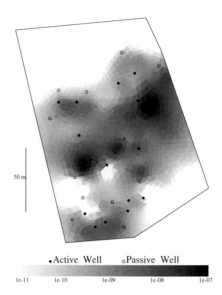

Figure 6.16: Kertess: *Dichlorethen* volatilization coefficient after calibration.

1. **Data:** In the application Kertess and for one given scenario (discharges of the active wells) we dispose of the NAPL concentration in the total extracted air which is assembled from all the wells.

2. **Parameters:** As for the transmissivity, using kriging, we build a volatilization coefficient field V_k from some measurements. We want to determine the positive constant c_v and to put
$$V = c_v V_k \,.$$

3. **Objective functional:** The constant c_v is taken as real scalar which minimizes the function
$$J_t(c_v) = (\hat{\eta} - \widetilde{\eta})^2 \tag{6.48}$$
with the notations : ($\widetilde{}$) means measured and ($\hat{}$) means computed.

6.5.5 Numerical Minimization Method

The two objective functionals (6.47) and (6.48) of this parameter identification are minimized using the genetic algorithm.

6.5.6 Results of the Calibration of the Volatilization Coefficient

Now, we present the the results of the calibration for: (A) dichlorethen, (B) tetrachlorethen, (C) trichlorethen and (D) the three NAPL components, see Figure 6.19. The total extraction rate is 240 $\frac{g}{s}$.

6.5. Parameter Identification of Kertess

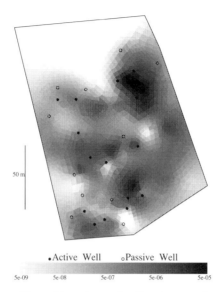

Figure 6.17: Kertess: *Tetrachlorethen* volatilization coefficient after calibration.

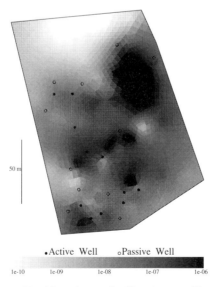

Figure 6.18: Kertess: *Trichlorethen* volatilization coefficient after calibration.

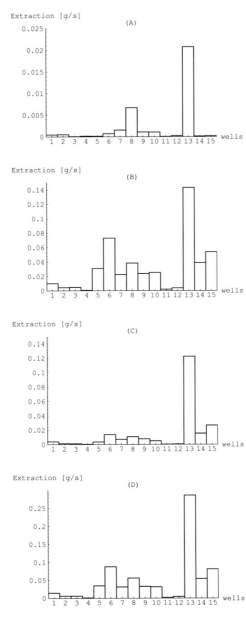

Figure 6.19: Kertess: Extraction rates with the original distribution of discharges and the total rate 240 $\frac{g}{s}$ for: (A) dichlorethen, (B) tetrachlorethen, (C) trichlorethen and (D) all three NAPL components.

Chapter 7

Numerical Methods and Optimization

Finite elements have had a great impact on the theory and practice of numerical methods during the twentieth century. There are dozens of monographs, thousands of papers devoted to their study. FEMs possess the properties which make them attractive for solving difficult physical and engineering problems. There exist various modifications of FEMs. Mixed finite elements represent a special family with a very important property – *the exact mass balance* Brezzi, Fortin [BF91] (the standard FEM do not have this). This is very important from the physical point of view. The idea of mixed finite elements is to approximate simultaneously the pressure (solution) and the velocity field in such a way that both converge in adequate norms to their continuous counterparts. The mixed approximation, in its original setting, leads to the resolution of indefinite algebraic systems, solution of which is quite delicate and time consuming. In recent years the popularization of mixed-hybrid formulation has increased rapidly. The reason is that the algebraic system associated with the original differential problem is positive definite.

We have chosen the mixed-hybrid formulation for space discretization by our computations. We explain briefly the main ideas of this approach in this chapter. We have used an unstructured multilevel grid by computer simulations. The grid generation is based on "GOOFE", i.e., a software tool developed by R. Hiptmair (University of Augsburg, Germany).

Although optimization-based approaches have been widely used in water remediation (cf. Gorelick [Gor90], Kuo, Michel and Gray [KMW92]), applications to SVE are rare (see Sun, Davert and Yeh [SMY96]). Most of the methods of nonlinear optimization are based on the existence of formulas for the derivatives of the cost functional. The nonexistence of these make the optimization more complicated. We describe some methods what to do in such cases.

The computations proceed in the following steps

1. solve the air flow equation (3.1) using the mixed-hybrid finite element method

2. compute the chemical transport equation (3.6)

3. optimize the extraction rates

7.1 Gas Flow Field

7.1.1 Linearity of the Flow Field w.r.t. the Control Variable

Before describing the finite element method, let us note that the air flow and its pressure are affine functions w.r.t. the control variable. Indeed, for $j = 1, \ldots, n$ let y_j and $\mathbf{q}_j$ denote the solutions of the BVP

$$\begin{cases} \nabla \cdot \mathbf{q}_j = -LTy_j - \delta_j & \text{in } G \setminus \overline{G_P} \\ \mathbf{q}_j = -T\nabla y_j & \text{in } G \setminus \overline{G_P} \\ \mathbf{q}_j \cdot \nu = lTy_j & \text{on } \Gamma_N \\ y_j = 0 & \text{on } \Gamma_D \cup \Gamma_P. \end{cases}$$

Let y_0 and $\mathbf{q}_0$ be the solution of the BVP

$$\begin{cases} \nabla \cdot \mathbf{q}_0 = LT(y_R - y_0) & \text{in } G \setminus \overline{G_P} \\ \mathbf{q}_0 = -T\nabla y_0 & \text{in } G \setminus \overline{G_P} \\ \mathbf{q}_0 \cdot \nu = -lT(y_R - y_0) + q_N & \text{on } \Gamma_N \\ y_0 = y_D & \text{on } \Gamma_D \cup \Gamma_P. \end{cases}$$

Then $y = y_0 + \sum_{j=1}^{n} u_j y_j$ and $\mathbf{q} = \mathbf{q}_0 + \sum_{j=1}^{n} u_j \mathbf{q}_j$ solve Problem 4.4.

Groups of Wells

In the case that the wells operate in groups (see the Remark at the end of Section 3.1.1) one first numerically determines the mapping $M : \mathbb{R}^m \to \mathbb{R}^m$ which maps the pressures y_k ($k = 1, \ldots, m$) onto the discharges u_k (see Section 4.1). Then one calculates the matrix corresponding to the inverse M^{-1}. This approach allows to determine the flow field belonging to any control vector $u = (u_1, \ldots, u_m)$.

Mixed Hybrid Finite Element Method

Let us consider the problem (4.4) which results from Problem 4.4 (the air flow with Dirac-type sinks) after subtracting the fundamental solutions from the exact solution, i.e., after "subtracting the singularities". This can be presented in the form of a general elliptic problem of a type

$$\begin{cases} \nabla \cdot \mathbf{q} + Cy = F & \text{in } G \\ \mathbf{q} + A\nabla y = 0 & \text{in } G \\ \mathbf{q} \cdot \nu = q_N & \text{on } \Gamma_n \\ y - \alpha \mathbf{q} \cdot \nu = -\beta & \text{on } \Gamma_d. \end{cases} \quad (7.1)$$

Here, the Γ_d consists of three parts: the part of Γ with a non zero leakage term, Γ_D and Γ_P. The rest of the boundary with a prescribed flux is denoted by Γ_n. The functions A, C, F, α, β and q_N are given and they depend only on the space variable. Moreover, for A and C we suppose that $0 < c_1 < A(\mathbf{x}) < c_2$ and $0 < c_3 < C(\mathbf{x}) < c_4$ for all $\mathbf{x}$

7.1. Gas Flow Field

in $\overline{G}$ (c_1, c_2, c_3 and c_4 are positive constants). Hence, $\nabla \cdot \mathbf{q} + Cy$ is a uniformly elliptic differential operator.

We use the mixed hybrid finite element method (MHFEM) for the space discretization. The original mixed method leads to an indefinite algebraic matrix. This difficulty can be removed by hybridization using Lagrange multipliers. For more details about the theoretical and numerical aspects of the MHFEM we refer the reader to Thomas [Tho77] and Brezzi, Fortin [BF91]. The reader can also find more details about the basis functions and methods for assembling the matrices in Kaasschieter, Huijben [KH90] from which we have taken most of the notations.

We now present the lowest order MHFEM to solve problem (7.1). We preserve the same notations for our variables when going from the PDE to the discrete form in order to avoid confusion.

7.1.2 Mixed Variational Formulation

Let us introduce the following notations

- $$L^2(G) = \{\psi : G \to \mathbb{R} | \int_G \psi^2 < \infty\}$$
  the Hilbert space with the norm
  $$\|\psi\|_{0,G} = (\int_G \psi^2)^{1/2},$$

- $$H(\text{div}; G) = \{\mathbf{v} \in (L^2(G))^2 | \nabla \cdot \mathbf{v} \in L^2(G)\}$$
  the Hilbert space with the norm
  $$\|\mathbf{v}\|_{\text{div},G} = (\|\mathbf{v}\|_{0,G}^2 + \|\nabla \cdot \mathbf{v}\|_{0,G}^2)^{1/2},$$

- the subspace of $H(\text{div}; G)$
  $$H_N(\text{div}; G) = \{\mathbf{v} \in H(\text{div}; G) | \mathbf{v} \cdot \boldsymbol{\nu} = 0 \text{ on } \Gamma_n\},$$

- and the variety, translation of $H_N(\text{div}; G)$
  $$H_*(\text{div}; G) = \{\mathbf{v} \in H(\text{div}; G) | \mathbf{v} \cdot \boldsymbol{\nu} = q_N \text{ on } \Gamma_n\}.$$

Then the mixed variational formulation of problem (7.1) is given by:

Find $(\mathbf{q}, y) \in H_*(\text{div}; G) \times L^2(G)$ such that $(\forall \mathbf{v} \in H_N(\text{div}; G), \forall \psi \in L^2(G))$

$$\begin{aligned}
\int_G (A^{-1}\mathbf{q}) \cdot \mathbf{v} + \int_{\Gamma_d} (\alpha \mathbf{q} \cdot \boldsymbol{\nu})(\mathbf{v} \cdot \boldsymbol{\nu}) - \int_G y \nabla \cdot \mathbf{v} &= \int_{\Gamma_d} \beta(\mathbf{v} \cdot \boldsymbol{\nu}) \\
\int_G \nabla \cdot \mathbf{q} \psi - \int_G Cy\psi &= -\int_G F\psi .
\end{aligned} \quad (7.2)$$

The existence and uniqueness of the solution of problem (7.2) is based on the "inf-sup condition" given by

$$\inf_{\psi \in L^2(G)\backslash 0} \sup_{\mathbf{v} \in H_N(\mathrm{div};G)\backslash 0} \frac{\int_G \psi \nabla \cdot \mathbf{v}}{\|\psi\|_{0,G} \|\mathbf{v}\|_{div,G}} \geq \gamma$$

where $\gamma > 0$ depends only on G.

7.1.3 Discrete Variational Formulation

We give here the discrete formulation which corresponds to the MHFEM. We define

- $\mathcal{T}_h$ the regular triangulation of the two dimensional bounded set G, more precisely G is supposed to be a domain with a polygonal boundary. $\mathcal{T}_h$ is then the set of triangles which form by their union $\bar{G}$ and satisfy the two following conditions
    - Conformity: Intersection of two different triangles is empty, one common edge or one common vertex.
    - Regularity: The minimal angle (taken over all triangles $\mathcal{T}$ in $\mathcal{T}_h$) is bounded from below by a positive constant.

- $E_h = \bigcup_{\mathcal{T} \in \mathcal{T}_h} \partial \mathcal{T}$,

- $RT^0(\mathcal{T}) = \{(a + bx_1, c + bx_2), a, b, c \in \mathbb{R}\} \subset (P_1(\mathcal{T}))^2$, where $\mathcal{T} \in \mathcal{T}_h$, $P_1(\mathcal{T})$ is the four dimensional space of the linear functions in x and y.

- $RT^0_{-1}(\mathcal{T}_h) = \{\phi \in (L^2(G))^2, \phi_{|\mathcal{T}} \in RT_0(\mathcal{T}), \forall \mathcal{T} \in \mathcal{T}_h\}$,

- $M^0_{-1}(\mathcal{T}_h)$ the space of piecewise constant functions on $\mathcal{T}_h$ (constant on each element),

- $M^0_{-1}(E_h)$ the space of piecewise constant functions on E_h (constant on each edge),

- $M^0_{-1,D}(E_h) = \{\lambda \in M^0_{-1}(E_h) | \lambda = 0 \text{ on } \Gamma_d\}$.

The hybrid version of the lowest order Raviart-Thomas mixed method for problem (7.2) is given by:

Find $(\mathbf{q}, y, \lambda) \in RT^0_{-1}(\mathcal{T}_h) \times M^0_{-1}(\mathcal{T}_h) \times M^0_{-1,D}(E_h)$ such that

$$\begin{aligned}
\int_G (A^{-1}\mathbf{q}) \cdot \mathbf{v} &+ \int_{\Gamma_d} (\alpha \mathbf{q} \cdot \boldsymbol{\nu})(\mathbf{v} \cdot \boldsymbol{\nu}) \\
- \sum_{\mathcal{T} \in \mathcal{T}_h} \left(\int_{\mathcal{T}} y \nabla \cdot \mathbf{v} - \int_{\partial \mathcal{T}} \lambda \mathbf{v} \cdot \boldsymbol{\nu}_{\mathcal{T}} \right) &= \int_{\Gamma_d} \beta \mathbf{v} \cdot \boldsymbol{\nu} \quad \forall \mathbf{v} \in RT^0_{-1}(\mathcal{T}_h) \\
\int_G \nabla \cdot \mathbf{q} \psi \quad - \int_G Cy\psi &= -\int_G F\psi \quad \forall \psi \in M^0_{-1}(\mathcal{T}_h) \\
\sum_{\mathcal{T} \in \mathcal{T}_h} \int_{\partial \mathcal{T}} (\mathbf{q} \cdot \boldsymbol{\nu}_{\mathcal{T}}) \mu &= \int_{\Gamma_n} q_N \mu \quad \forall \mu \in M^0_{-1,D}(E_h)
\end{aligned} \quad (7.3)$$

7.2. Contaminant Transport

Let us choose the bases for the spaces $RT^0_{-1}(\mathcal{T}_h)$, $M^0_{-1}(\mathcal{T}_h)$ and $M^0_{-1,D}(E_h)$ as described in Kaasschieter and Huijben [KH90]. The discrete variational formulation (7.3) leads to a linear system of the form

$$\begin{cases} A\mathbf{q} + B^t y + L^t \lambda = G \\ B\mathbf{q} - Cy = F \\ L\mathbf{q} = H \, . \end{cases} \quad (7.4)$$

Solving System (7.4) We want to determine the unknowns $\mathbf{q}, y, \lambda$ in (7.4). First, using the formal elimination, we compute λ, then y and at the end $\mathbf{q}$. We describe briefly all steps (cf. Kaasschieter and Huijben [KH90]). Let us note that the matrix A is block diagonal. Each block is 3×3 and positive definite. Thus its inverse A^{-1} can be easily computed.

- Solve the system

$$[L[Id_A - MB]A^{-1}L^t]\lambda = L[-MB + Id_A]A^{-1}G + LMF - H$$

for λ, where $M = A^{-1}B^t[B\,A^{-1}B^t + C]^{-1}$. The matrix $L[Id_A - MB]A^{-1}L^t$ is sparse, symmetric, and positive definite. One can effectively solve it using a preconditioned conjugate gradient method.

- Solve the system

$$[BA^{-1}B^t + C]y = (-F + BA^{-1}G - B\,A^{-1}L^t\lambda)$$

for y. Now $[BA^{-1}B^t + C]$ is a positive definite diagonal matrix which makes its inversion trivial.

- After getting λ and y solve

$$\mathbf{q} = A^{-1}[G - B^t y - L^t \lambda]$$

for $\mathbf{q}$.

7.2 Contaminant Transport

Let us note that the dissemination of NAPLs by diffusion or dispersion is very small with respect to the convection caused by extraction wells. This makes the numerics more complicated. Taking into account the diffusion ($D > 0$), we can use a finite element method adapted to convection dominated flow problems. In the case when diffusion is negligible with respect to convection, the streamline method can be used for computations. We will concentrate our attention to the last case.

Application of the Streamlines Method to Flow Fields Given by a MFE Discretization

We first want to compute a streamline (in backward direction) on a given triangle $\mathcal{T}$ of the triangulation. According to the fact that the bases functions for flux approximation are piecewise linear, it is easy to see that the streamlines are straight lines within each triangle. Indeed, from the definition of $RT^0(\mathcal{T})$ in Section 7.1.1 the flux vector $\mathbf{q}$, and therefore, any right-hand side of the differential equation (3.46), is a linear function

$$\mathbf{q} = \sum_i \alpha_i(\mathbf{x} - \mathbf{x}_i) = \sum_i \alpha_i \mathbf{x} - \sum_i \alpha_i \mathbf{x}_i,$$

i.e., $\mathbf{q} = \lambda(\mathbf{x} - \mathbf{x}_C)$ with $\lambda \in \mathbb{R}$ and $\mathbf{x}_C \in \mathbb{R}^2$ or $\mathbf{q} = -\mathbf{x}_C$. Therefore, any solution of (3.46) is of the form

$$\mathbf{x}(t) = \mathbf{x}_C + (\mathbf{x}_0 - \mathbf{x}_C)e^{\lambda t} \text{ or } \mathbf{x}(t) = \mathbf{x}_0 - \mathbf{x}_C t.$$

Therefore, the streamlines are straight lines inside each triangle.

We assume that $\mathcal{T}$ has the vertices $\mathbf{x}_1, \mathbf{x}_2, \mathbf{x}_3$ and edges e_1, e_2, e_3 with e_j opposite to $\mathbf{x}_j$. If we consider the streamline leaving the triangle at the point $\mathbf{x}'_0$ on e_j (see Figure 7.1), we first compute the vector product components

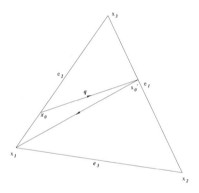

Figure 7.1: Streamline segment in a finite element.

$$\alpha_0 = (-\mathbf{q} \wedge (\mathbf{x}_j - \mathbf{x}'_0))_{x_3} = q_{x_1}(\mathbf{x}'_0 - \mathbf{x}_j)_{x_2} - q_{x_2}(\mathbf{x}'_0 - \mathbf{x}_j)_{x_1} \quad (7.5)$$
$$\alpha_i = ((\mathbf{x}'_0 - \mathbf{x}_j) \wedge (\mathbf{x}_i - \mathbf{x}_j))_{x_3} \quad (7.6)$$

for $i \neq j$. If now $\text{sign}(\alpha_0) \neq \text{sign}(\alpha_i)$, then the point $\mathbf{x}_0$ where the streamline enters $\mathcal{T}$ is on the edge e_i, otherwise it is not. (If x'_0 coincides with one of the vertices or if the flux $\mathbf{q}$ is parallel to $\mathbf{x}_1 - e_1$, then a small modification of the formulas is in order.) We are now able to compute the second endpoint $\mathbf{x}_0$ of the streamline. If $\mathbf{x}_0$ is on e_i and the third edge of the triangle is e_k we get

$$\mathbf{x}_0 = c_1 \mathbf{x}_j + (1 - c_1)\mathbf{x}_k = \mathbf{x}'_0 - c_2 \mathbf{q} \quad (7.7)$$

which gives the linear system

$$c_1(\mathbf{x}_j - \mathbf{x}_k) + c_2 \mathbf{q} = \mathbf{x}'_0 - \mathbf{x}_k \quad (7.8)$$

for the coefficients c_1, c_2. We take the line parameter t such that $\mathbf{x}(t) = \mathbf{x}_0 - t\mathbf{q}$. If we assume that the transfer coefficient $V(\mathbf{x})$ and the leakage term $L(\mathbf{x})T(\mathbf{x})(y_R - y)$ are affine functions along each line segment, according to formula (3.47) we have to solve an initial value problem

$$\frac{d\eta(t)}{dt} = a_0(t)\eta(t) + a_1(t) , \quad \eta(0) = \eta_0$$

on this line segment where a_0 and a_1 are affine functions of t. The solution of this problem is

$$\eta(t) = \left(\eta_0 + \int_0^t a_1(s)e^{-A_0(s)}\, ds\right) e^{A_0(t)} \text{ with } A_0(t) = \int_0^t a_0(s)\, ds. \quad (7.9)$$

The relative concentration at $\mathbf{x}'_0$ is

$$\eta(t_0) \text{ with } t_0 = \frac{|\mathbf{x}'_0 - \mathbf{x}_0|}{|\mathbf{q}|}.$$

7.3 Integrating the Extraction Rate

The NAPL concentration η computed in Section 7.2 is used to evaluate the quantity of NAPLs extracted by each active well (as defined by the objective function (3.7)). The distribution of η around each well is highly oscillatory. The reason is that some streamlines ending by an extraction well cross the high contaminated subregions on their way to the probe and others do not, i.e., the nonuniform distribution of the contaminant in the remediation site causes the oscillations of η along the active wells. Then it is natural to think about using an adaptive quadrature integration rule to approximate J with a good accuracy and a reasonable number of evaluations of η. The adaptive quadrature we have used can be presented as follows (for more details see Maron [Mar87]):
Suppose that we are computing the integral

$$\int_a^b f(t)dt$$

where a and b are two real numbers and f is a function defined on the interval $[a, b]$ but not necessarily continuous. We make use of Simpson's rule

$$S_a^b = \frac{b-a}{6}[f(a) + 4f(m) + f(b)]$$

with $m = \frac{a+b}{2}$ which is of order $O(h^4)$ with $h = b - a$. We say that the integration is accurate enough if

$$S_a^m + S_m^b - S_a^b < \varepsilon$$

otherwise we do the same with the two parts of $[a, b]$, namely $[a, m]$ and $[m, b]$ separately.

7.4 Optimization Algorithm

Most of the nonlinear programming optimization techniques require the derivatives of the cost functional (cf. Lions [Lio71]). Unfortunately, no explicit forms of these are available in many real cases. Therefore, a perturbation method is needed to generate the system Jacobian matrix which approximates gradients. In this section we describe a suitable method for optimization without the derivatives of J.

Characteristics of the Optimal Control Problem

Our optimization problem is to maximize the quantity of extracted chemicals for a given total pumping rate which is the same as to minimize the functional

$$J(u_1,\ldots,u_n) = -\int_{\Gamma_A} \mathbf{F}\cdot\boldsymbol{\nu}\, d\Gamma \qquad (7.10)$$

subject to the two constraints

- $\sum_{j=1}^{n} u_j = u_T$ (limited total pumping rate)

- $u_j \geq 0$ for $j = 1,\ldots,n$.

We can summarize the characteristics of the optimal control problem as

1. The functional J must be negative and bounded from below

2. We consider J without its derivatives. The streamlines method which is used to compute the extraction rate and the NAPL concentration along some lines drown by the flow field does not offer the possibility to compute the gradient of J.

3. The dimension of the control variable is less then 20

4. J could have multiple local minimum. If the pumping procedure uses only one active well, one can see that the flux in the domain and the quantity of extracted NAPLs are monotone functions of the discharge of this well (see Chapter 5 Figures 5.2 and 5.4). Contrary if more than one well act at the same time the situation changes radically and it is hard to predict without numerical tools an optimal distribution of the discharges to the wells. This complexity grows with the number of the wells.

To deal with this problem we perform two steps. In the first step we apply a global optimization namely genetic algorithm to look for all potential global minimum in the whole domain. In the second step we use a local search method to improve the best(s) minimum found in the first step. We have chosen Powell's algorithm as local search which is adapted for minimization of functionals without derivatives. It is possible to use

7.4. Optimization Algorithm

descent methods like gradient or Newton methods where the derivatives are computed by differentiating the objective functional numerically

$$\frac{\partial J}{\partial u_i} \approx \frac{J(u_i + h) - J(u_i)}{h}$$

where h is a small real number. Such a technics reduce the effectiveness of these descent methods, therefore we avoid this. The two methods, genetic algorithm and Powell's algorithm, are given here in more detail.

Genetic Algorithm

The genetic algorithm we present here is used to maximize positive functional. If one has to deal with minimization or with negative functional or both, it is easy to transform the problem using some trivial linear mapping. We give here the principal steps of this algorithm and for the details about its origin, development and mathematical foundations we refer to Goldberg [Gol89]. A genetic algorithm could be written in different ways and depend on many parameters but they all look more or less like the following simple iteration.

1. Data structures: The set of the control variable is encoded using the binary set $X = \{0, 1\}^M$. $card(X) = 2^M$ and each element of X can be decoded to one control variable. The objective functional is called fitness function $f : X \longrightarrow \mathbb{R}_+$. We denote $\mathcal{P}$ any element of X^N, we call it population, N is one fixed integer and it is not necessary that all the components of $\mathcal{P}$ are different. Sometimes, one component of $\mathcal{P}$ is called element or individual. In this algorithm, we mean by randomly the uniform distribution only if another distribution is mentioned.

2. Initialization: N elements are taken randomly in X and constitute the initial population $\mathcal{P}_0$.

3. For given population $\mathcal{P}_t$, the genetic algorithm generates a new population $\mathcal{P}_{t+1}$ using three transformations; $\mathcal{P}_{t+1} = T_M T_C T_R(\mathcal{P}_t)$ where

$$T_M, T_C, T_R : X^N \longrightarrow X^N$$

and are defined by

(a) T_R is called *reproduction*. For given population $\mathcal{P} = (x_1, \ldots, x_N)$ we evaluate the fitness function f for all $x_i, i = 1, \ldots, N$. The N individuals $y_j, j = 1, \ldots, N$ of $T_R(\mathcal{P})$ are then generated randomly in $\{x_1, \ldots, x_N\}$ following the distribution

$$P_r\{y_j = x_i\} = \frac{f(x_i)}{\sum\limits_{k=1}^{N} f(x_k)}.$$

In this first transformation, the elements in the population are copied with probability which corresponds to their objective function value. Obviously, this will produce many copies of the best elements.

(b) T_C is the *crossover* operator. We repeat for L_c times: {select randomly x and y two individuals (components) of $\mathcal{P}$. The two elements x and y are in X and can be written in the binary form $x = x_1 x_2 \ldots x_M$ and $y = y_1 y_2 \ldots y_M$. Select randomly an integer k between 1 and M and form the two new individuals $x' = x_1 x_2 \ldots x_{k-1} y_2 \ldots y_M$ and $y' = y_1 y_2 \ldots y_{k-1} x_k \ldots x_M$. x' and y' are called children and they will replace their parents x and y in the new population $T_C(\mathcal{P})$}.

(c) By T_M we denote the *mutation* operator. For each individual x of $\mathcal{P}$ with probability p_m we do {select randomly an integer k between 1 and M and flip the k'th bit x_k to produce the new individual in $T_M(\mathcal{P})$}.

4. If no STOP criterion is met then GOTO step 3.

Remark 7.1 As any other method, the genetic algorithm has many parameters which have to be specified.

1. Mapping the objective functional which can affect the speed of convergence.

2. The length M of the individuals. If the original control variables lie in the vectorial space $\mathbb{R}^n$ then each time M is big, one can encode more of its elements to the set X.

3. The size N of each population which is taken here constant. One can also consider populations with different sizes as nature would do.

4. The number L_c of the *crossover* and the probability p_m of the mutation. The general tendency is to take a height number for the crossover and a low mutation probability ($p_m \approx 1/N$)

5. The stopping rule could be the number of the iterations t, the convergence which means no improvement in the fitness function or a similarity in the populations.
□

Powell's Algorithm

The genetic algorithm could lead to one global or many similar local minimizers of J. Near each of these minimizers, we use Powell's algorithm to perform local search or to improve the minimum of J. This does not require the derivatives of the functional J and is given by the following iterative procedure (cf. Brent [Bre73]). First we eliminate the constraint by redefining the objective function J. We introduce $m = n - 1$ and $w = (w_1, \ldots, w_m) \in \mathbb{R}^m$ and use

$$\tilde{J}(w) = J\left(w_1, \ldots, w_m, u_T - \sum_{j=1}^{m} w_j\right)$$

1. w_0 an initial approximation of the minimizer

2. set $v_1, \ldots, v_m$ as the columns of the identity matrix

3. for $j = 1, \ldots, m$ compute β_j the minimizer of $\tilde{J}(w_{j-1} + \beta_j v_j)$ and put $w_j = w_{j-1} + \beta_j v_j$

4. for $j = 1, \ldots, m-1$ replace v_j by v_{j+1}

5. replace v_m by $w_m - w_0$

6. compute β to minimize $\tilde{J}(w_0 + \beta v_m)$ and replace w_0 by $w_0 + \beta v_m$

7. If the convergence is achieved then stop otherwise go to step 3.

The steps $3, \ldots, 7$ represent one iteration in Powell's algorithm.

This algorithm has two difficulties which have to be considered.

- The minimization of the one variable function in steps (3) and (6) of the iteration. If we assume that we are close enough to the minimizer of J then J is locally convex and we can approximate it by a quadratic polynom using its value on three different points. For more details see Brent [Bre73].

- The linear dependence; it is observed that the directions $v_1, \ldots, v_m$ sometimes become dependent and then the algorithm looks for a minimum in proper subspaces of $\mathbb{R}^m$. The simplest way to avoid linear dependence of the search directions and retain quadrature convergence is to reset the search directions $v_1, \ldots, v_m$ to the columns of the identity matrix after every m or $m+1$ iterations Brent [Bre73].

7.5 Programming Aspects

The computation of the flow field has the following aspects

- The computational domain G is a 2-D bounded domain with a polygonal boundary (see Sections 8.1, 8.2).

- The passive wells are represented by small holes inside G. The atmospheric pressure is prescribed on Γ_P (the boundary of passive probes) as the Dirichlet boundary condition.

- For active wells we have considered two cases

    - Active wells are modeled as the point sinks using Dirac functions. In this case the extraction probes together with their vicinities belong to G (see Section 8.1).

– Active wells are modeled by a given differential pressure (Dirichlet boundary condition). In this event the extraction probes do not belong to G, thus they are cut of from the remediation site together with small vicinities (see the application in Section 8.2).

- We use an unstructured triangular mesh with local refinement near all wells in order to approximate better the singular behavior of solution at the wells. The number of triangles was at about 2.E04 triangles. For mesh generation we have used "GOOFE"- the software tool created by R. Hiptmair (University of Augsburg, Germany). Here the RED-GREEN algorithm for refinement of triangles is used.

- The program is written in C/C++ language and run on a SUN SPARC-workstation 5 machine.

Chapter 8

Applications

The methodology described in the previous chapters was applied to two representative soil remediation sites in northern Germany where the soil venting technique is being used to clean-up the water-unsaturated soils contaminated by volatile organic compounds among other contaminants. Model testing and application was carried out in close cooperation with an engineering company that is responsible for the practical soil venting operation and VOC-clean-up. The data base provided by the company represents the type of relatively limited information about the problem and the soil properties that are usually available for most practical soil venting operations.

Typical remediation sites have been selected where soil venting is not only induced by a single extraction well but by a number of venting wells operating on a larger contaminated area, such that vapor extraction at neighboring wells may be influenced by the air extraction of other wells and where the effectiveness of the whole venting installation has to be optimized.

The first test site (Kertess), which was already in operation when the project started, was used for the development of the numerical flow model, the optimization model, and parameter estimation procedure while the second site (Kirchweyhe) was intended to test the derived optimized configurations in a real case field application.

8.1 Kertess

The Kertess test-site is an abandoned 2 ha large industrial area in Hannover, Germany which has been contaminated by approximately 100 tons of VOC's (among other contaminants). The test-site contains several buildings including basements and abandoned railway tracks (Figure 6.6). Large portions of the soil surface are sealed by asphalt or concrete, however, the 5-10 cm thick sealing layer is partly destroyed or may contain cracks and fissures. The following data have been available:

Qualitative soil descriptions (soil protocols) from about 100 borehole locations (Figure 6.6).

Contaminant concentration measurements in the soil-air at several locations.

Air-permeability measurements at 14 undisturbed soil cores of 100 cm^3 volume sampled from a medium sand of the originally layered material from a depth of about $3 - 3.6\ m$ (unpublished report by A. Lange, Institute of Water Management, Hydrology and Agricultural Hydraulic Engineering, University of Hannover, Germany).

Air discharge and air pressure measurements at the installed 16 extraction (active) and 12 air-inlet (passive) wells (provided by GEO-data).

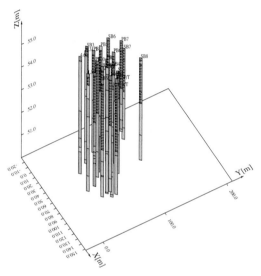

Figure 8.1: Kertess: Soil textures and layers above the ground water table of some borehole locations obtained from soil protocols.

Figure 8.1 gives an impression of the results of the soil profile descriptions. The soil textures for each layer above the ground water of all borehole locations and the elevations is shown in a quasi-three-dimensional image. Note, that the vertical axis is enlarged about 40 times compared to the two horizontal axes. The ground water table is about 5 m below surface. For the location of the boreholes see Figure 6.6. Figure 8.1 shows that in the upper part of most profiles underneath the asphalt layer the soil consists of anthropogenetically disturbed and refilled material consisting mostly of a sandy soil matrix which contains varying amounts of gravel, debris or rubble (gray colors with black dots in Figure 8.1). Originally layered soils can be found only in the lower parts of the profiles where fine, medium, or coarse textured sandy materials are dominating.

Figures 6.15 and 6.18 give an impression of the spatial distribution of the VOC-component Trichlorethen (TCE) at the 2 ha test-site.

Soil-air samples have been taken from several locations and the contaminant concentration has been analyzed. The data have been used to draw maps of the contaminant distribution over the Kertess-area. Spatial interpolation of the contaminant concentra-

8.1. Kertess

	measured mean (range)	estimated mS1, mS2
specific permeability $[10^{-12} m^2]$	15 (5-25)	3, 10
porosity $[m^3/m^3]$	0.4-0.48	0.4, 0.43

Table 8.1: Kertess: Comparison between estimated (Table 6.6) and measured air permeabilities and porosities for a medium sand. The measured values are obtained using $100 cm^3$ volume soil cores (cf. Lange [Lan92]).

tion has first been done empirically by GEO-data. The volatilization coefficient, V, shown in Figure 6.18 is assumed to be linearly related to the liquid phase concentration and thus, reflecting areas with relatively high (black) and low (white) contaminations. The TCE-content has been integrated over the uppermost 3–4 m of the soil. The TCE-map in Figure 6.18 is derived from measured data of the TCE-density in the soil air at specific locations (cf. Figure 6.6) which have been spatially interpolated using kriging (see Section 6.3).

At the Kertess test-site, 12 air supply (passive) and 16 extraction (active) wells have been installed (Figure 6.7). The location of the wells has been selected empirically according to the initial survey of contaminant and soil property distribution. The water-saturated zone of the test-site is completely isolated horizontally by a bung-wall in order to prevent contaminant migration within the ground water zone towards the surrounding systems. Control measurements at observations wells indicated that the water table remained fairly constant at a depth of about 5 m below soil surface. Flow of air through the water-unsaturated upper part of the soil is not restricted by the side-walls.

Since the depth of the aerated zone (5 m) is small compared to the horizontal extent of the contaminated area (approximately 150 m X 200 m), we assume an essentially two-dimensional (2-D) horizontal model to describe the movement of air soil. However, the horizontal-flow assumption may not strictly hold because of impermeable objects in the ground, such as foundations of buildings or concrete ditches, which may serve as partial "barriers" and force the air flow field to deviate to some extent. In order to keep the model development as simple as possible and since detailed information is not available, we neglect the three-dimensional effects in this study.

For estimating the transmissivities of the 2 ha test-site, the permeabilities at each location and of each of the layers are estimated following the procedure described in Chapter 6. The parameter estimation procedure begins with a classification of the porous materials into A) original layered, B) refilled, and C) rubble material. The classification depends on the relative content of stones or gravel described qualitatively in the soil protocols. Soil physical properties are estimated using "pedotransfer functions" applied to the "fine" soil material (particle-sizes smaller than 2 mm when considering an "equivalent" particle diameter). The relative proportion of the stone content is then used to modify the estimates of the porosity and the tortuosity coefficient (for details see Chapter 6). The Kozeny-Carman theory seems to provide relatively good estimates of the specific permeability for the sandy soils (see Table 8.1).

Estimated values for a medium textured, originally layered sandy soil are in the same range compared to measured data. Since no measured data for the parameters of the other materials are available, results of the air-flow simulations will be calibrated using the discharge and pressure measurements (see below).

To obtain pressure-saturation relations we used the "pedotransfer"-function of Haverkamp and Parlange [1986] and fitted the results to the van Genuchten-type retention and relative permeability function. The resulting parameters are then used also for the two-phase (air-liquid) system by applying the theoretical approach proposed by Parker et al. [PLK87]. Pressure-saturation functions are used to calculate the air-filled porosity as a function of depth for each of the borehole descriptions assuming a linear pressure head depth-profile between the ground water table and the soil surface. The linear pressure head profile represents hydrostatic conditions which may be valid for the site-specific conditions for the Kertess area where we have a constant water table and a sealed surface. Air permeabilities as a function of depth are obtained at each borehole location by using the relative permeability function and the phase-filled porosity and by using the relation between the dynamic viscosity of the liquid and the gas phase. Integration of the air permeability in vertical direction between the ground-water table and the surface results in the transmissivities for each borehole location. In a similar way, the source term, f, representing the air extraction by venting wells, was integrated over depth. Spatial interpolation of T and F in the 2-D-horizontal plane was obtained using kriging (see Chapter 6). The resulting spatially interpolated transmissivity field is shown in Figure 6.9. These areas mainly reflect the location of foundations of the buildings. For basements and foundations we assumed a zero air-permeability, so that only the narrow and low permeable aerated lower part of the soil profile may allow some air movement underneath the buildings. In general, the transmissivity values are dominated by the permeability of the more aerated upper parts of the profiles where refilled and rubble material is dominating.

Results of the numerical air-flow simulations using the estimated transmissivity field are given in Figures 6.11 and 6.12. The results are presented for a) the 2-D-air pressure field and b) the flow velocities as influenced by the pumping of air at the 16 extraction wells (acting as sinks) and the 12 passive wells (acting as sources).

Since the upper 1–2 meters of the soils consist of refilled, gravel, or rubble material, the uncertainty of the transmissivity estimation is evident. The transmissivity field was calibrated to improve the parameter estimation by fitting the calculated discharges at the 12 passive wells to the measured values such that the average discharge of all the wells agreed with the measured ones. For the refilled and rubble material, which is a non-soil-type porous medium, thoroughly tested pedotransfer functions (i.e., empirical estimations) are hardly available.

The air pressure field (Figure 6.11) and the flow velocity field (Figure 6.12) were calculated using the improved transmissivity distribution show that relatively small areas containing rubble and gravel lenses make up for the large air velocities and discharges. Further improvement could easily be obtained when calibrating the transmissivities locally for each of the extraction wells (which was not done here).

8.1. Kertess

The conclusions of the parameter estimation and calibration procedure in case 1 are:

1. The approach may be used to simulate soil-air flow fields for relatively large contaminated areas with sandy soils utilizing limited data and qualitative soil profile information which are often only available in practical venting operations.
2. The assumption of two-dimensional, steady horizontal air-flow may be critical considering the occurence of cracks and fissures in the asphalt layer and in the rubble material as well as the effect of foundations of buildings on the flow field.
3. The parameter estimation procedure using pedotransfer functions and empirical relations gives reasonable results for original layered sandy soil materials.
4. Predictions of parameters for mechanically disordered and non-soil-type material, such as rubble, gravel, rock and wood debris, is highly uncertain.
5. Soil physical properties of such waste material can be extremely variable and should be studied in more detail, since soil contaminations often occur at industrial sites.

8.1.1 Optimizing the Extraction Rates

The cleaning procedure uses 15 active and 11 passive wells given at the fixed positions shown in Figure 6.13. The parameters of the flow and transport equations are as described in Section 6.5. In the same section, Figure 6.19 shows the extraction rates of the pollutants.

The optimization of the extraction rates is performed in the following steps:

Optimization of discharges with different total rates. The aim of this step was to optimize the discharge of single wells using the maximum pollutant extraction as the objective functional with the constraint to keep the total air extraction constant. We applied this procedure to three different chemicals namely *cis-1,2-Dichlorethen, Tetrachlorethen, Trichlorethen* first separately and then *for the sum of all three*. Extraction rates have been optimized for the three different air extraction rates of 240, 100 and 10 $\frac{g}{s}$.

The optimized air discharges for a total discharge rate of 240 $\frac{g}{s}$ when extracting *cis-1,2-Dichlorethen, Trichlorethen, Tetrachlorethen*, and when considering the sum of the three components are given in Figure 8.3 (A), (B), (C), and (D), respectively. The pollutant extraction rates corresponding to the air discharges are given in Figure 8.4. Figures 8.5 and 8.6 show the optimized air discharges and pollutant extraction rates for the total discharge rates of 100 $\frac{g}{s}$ and the results for discharge rate of 10 $\frac{g}{s}$ is represented by Figures 8.7 and 8.8. The different extraction rates are caused by the nonuniform distribution of pollutant in the soil matrix.

After summing up the pollutant extraction rates of the single wells the total extraction is shown in Figure 8.9 for the original distribution (A) and the optimized cases with total discharge rates of 240 (B), 100 (C) and 10 (D) $\frac{g}{s}$. The values represented in Figure 8.9 are given in Table 8.2 and Figure 8.2. In other words the optimal discharges

	Total Air Discharge Rate			
contaminant	Orig. distribution $(240\ \frac{g}{s})$	Optim. A $(240\ \frac{g}{s})$	Optim. B $(100\ \frac{g}{s})$	Optim. C $(10\ \frac{g}{s})$
cis-1,2-Dichlorethen	0.0339	0.0348	0.0340	0.0327
Tetrachlorethen	0.477	0.540	0.451	0.138
Trichlorethen	0.220	0.230	0.222	0.149
All components	0.731	0.797	0.701	0.316

Table 8.2: Kertess: Pollutant extraction rates in g/s obtained with the original distribution used by GEO-data compared with the optimized scenarios for total discharge rates of 240, 100, and 10 $\frac{g}{s}$ for sum of three major contaminants and each separately. The values for all components represent the optimized extraction rate considering all the three components simultaneously except for the original distribution where it is the sum of the components above.

which give high extraction for *Trichlorethen* is not the optimum in the case of *Tetrachlorethen*: At different discharge rates the relation of the optimized extraction rate for the 3 components is changing perhaps because of the different volatilization coefficients which are used.

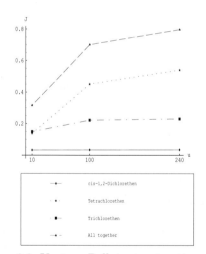

Figure 8.2: Kertess: Pollutant extraction rates.

It is obvious that with increasing total discharge the extraction rate is increasing, too. The monotonic behavior of the pollutant extraction with respect to the air flow field is caused by linear relations assumed for the volatilization of pollutants.

Compared with the two other components the extraction rate of *cis-1,2-Dichlorethen* is relatively constant: The extraction decreases about 6 % when changing the air

8.1. Kertess

discharge rate from 240 $\frac{g}{s}$ to 10 $\frac{g}{s}$. The explanation can be derived from the Figures 5.4 and 5.6 which show that for dimensionless total discharges larger than 100, there is hardly any variation of the extraction rate. The change from liquid to gaseous phase for the component *cis-1,2-Dichlorethen* is slow compared to the air flow velocities. At the lowest tested air discharge rate of 10 $\frac{g}{s}$ this contaminant can be removed as good as at higher flow rates.

The second row which corresponds to the component *Tetrachlorethen* shows a much different behavior and a big response of the pollutant extraction to the total air discharges. Optimization improves the extraction rate and the default value of 0.477 increases with 13.2 % to become 0.54 when increasing air discharges at wells where the concentrations are relatively high. By reducing the total air discharge rate to 100 and 10 $\frac{g}{s}$ the extraction rate decreases by 16.5 and 74.4 %, respectively. The relatively large reduction of the *Tetrachlorethen* extraction rate is due to the fact that the volatilization coefficient is almost 63 times larger than the one of *cis-1,2-Dichlorethen*. The spatial distributions of the volatilization coefficients of the three components are shown in Figures 6.16, 6.17 and 6.18.

The results in Table 8.2 and Figure 8.2 for the component *Trichlorethen* show a behavior which is somewhat in between the other two; its volatilization coefficient is 4.4 times smaller than the one of *Tetrachlorethen*. For *Trichlorethen* if we reduce the total discharge rate from 240 $\frac{g}{s}$ the extraction decreases with 3.5 % at 100 and with 33 % at 10$\frac{g}{s}$. These results suggest that a further increase of air pumping rates would lead to a stationary extraction while a further decrease of the air discharge would lead to a decrease of the extraction rates and they at the end exponentially approach zero.

Also, considering the concentration of the pollutant in the soil air at complete saturation for *Trichlorethen* is 2.5 larger than the one for *Tetrachlorethen*, thus the volume extraction of *Tetrachlorethen* at the relatively low air discharge rate (10 $\frac{g}{s}$) is smaller than that of *Trichlorethen*.

The spatial distributions of the optimized solutions for *Tetrachlorethen* extraction for air discharge rates of 100 $\frac{g}{s}$ are shown in Figure 8.10 and for 10 $\frac{g}{s}$ in Figure 8.11 together with some of the streamlines and a boxes representing the pollutant extraction volume of single active wells. The horizontal length of the box represents the discharge rate while the vertical one shows the average concentration of the air removed by the active well. The results show, that some extraction wells, e.g., A3, A4, are highly inefficient while others, A6, A8, A9, A13, and A15 have large extraction rates.

Varying the positions. In this part we vary locations of the wells and then optimize the discharges. The optimal choice of positions is a mathematical problem still to be solved. But as a first attempt one can place the wells at those spots where the concentration of the pollutant is large. In other words, one moves the pumps to the strongest sources of the NAPLs. The new 15 positions are taken as the 15 measurement points where the concentration is found the largest. This positions differ from one pollutant to another and since the quantity of extracted pollutant is also different it is not obvious to choose the best positions to optimize the total extraction of all the pollutants. In this part we have chosen to consider only the Tetrachlorethen because of its high extraction rate and

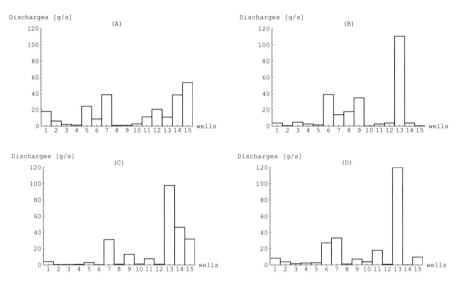

Figure 8.3: Kertess: Optimal discharges [$\frac{g}{s}$] of active wells with the total discharge 240 $\frac{g}{s}$ for the extraction of: (A) dichlorethen, (B) tetrachlorethen, (C) trichlorethen and (D) the three NAPL components.

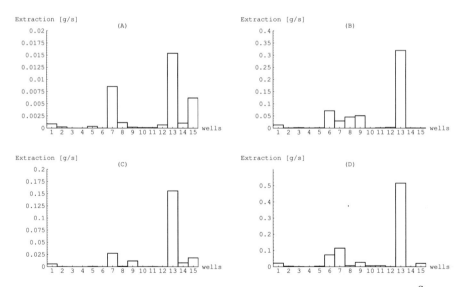

Figure 8.4: Kertess: Optimal extraction rates with the total discharge 240 $\frac{g}{s}$ for: (A) dichlorethen, (B) tetrachlorethen, (C) trichlorethen and (D) the three NAPL components.

8.1. Kertess

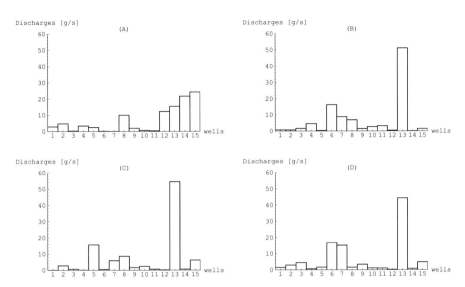

Figure 8.5: Kertess: Optimal discharges $[\frac{g}{s}]$ of active wells with the total discharge 100 $\frac{g}{s}$ for the extraction of: (A) dichlorethen, (B) tetrachlorethen, (C) trichlorethen and (D) the three NAPL components.

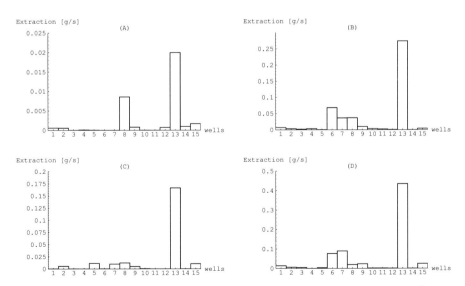

Figure 8.6: Kertess: Optimal extraction rates with the total discharge 100 $\frac{g}{s}$ for: (A) dichlorethen, (B) tetrachlorethen, (C) trichlorethen and (D) the three NAPL components.

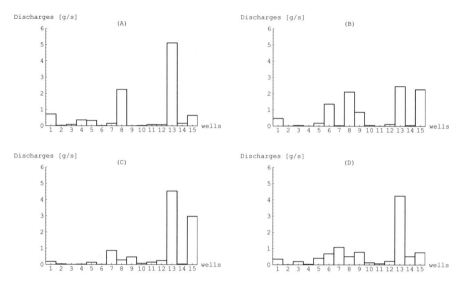

Figure 8.7: Kertess: Optimal discharges [$\frac{g}{s}$] of active wells with the total discharge 10 $\frac{g}{s}$ for the extraction of: (A) dichlorethen, (B) tetrachlorethen, (C) trichlorethen and (D) the three NAPL components.

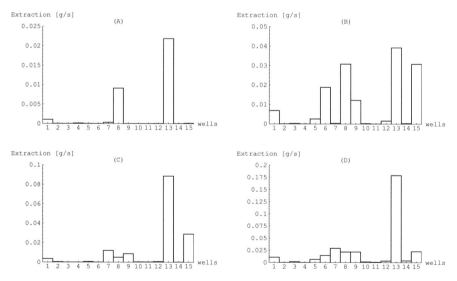

Figure 8.8: Kertess: Optimal extraction rates with the total discharge 10 $\frac{g}{s}$ for: (A) dichlorethen, (B) tetrachlorethen, (C) trichlorethen and (D) the three NAPL components.

8.2. Kirchweyhe

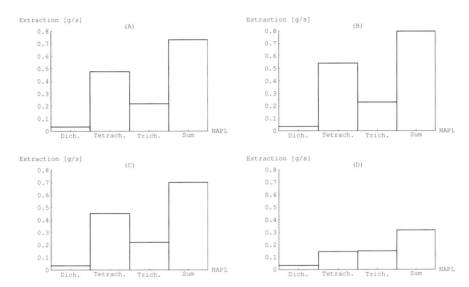

Figure 8.9: Kertess: Extraction rates for: (A) Original distribution, (B) Optimal with total of 240 $\frac{g}{s}$, (C) Optimal with total of 100 $\frac{g}{s}$ and (D) Optimal with total of 10 $\frac{g}{s}$ ($1\frac{g}{s} \approx 0.6\frac{Ton}{Week}$).

therefore the 15 wells are placed at the 15 highest measurement points. We optimize the Tetrachlorethen extraction with total air pumping rate of 100 and 10 $\frac{g}{s}$ respectively. We represent in Figure 8.12 the optimal discharges and the corresponding pollutant extraction which result from this varying the positions. The extraction using 100 $\frac{g}{s}$ is improved only by 3 % but it is improved 14.2 % when acting with 10 $\frac{g}{s}$. To explain this we say that the original positions may be also chosen using some other techniques or intuitions or knowledge about the origin of the pollutant and for a high air pumping rate the difference due to the new position is rather small. In the case of a pumping rate of 10 $\frac{g}{s}$ this difference is rather large and therefore with a leakage surface the positions of the wells plays an important role.

8.2 Kirchweyhe

The Kirchweyhe test-site is an industrial area of about $100m \times 150m$ large located near the city of Bremen in Germany. The ground water table is $3.2m$ under the soil surface, which is covered by concrete and sealed by iron plates in the central part of the test-site. The pedo- and geological situation is somehow similar to that of the Kertess-area with respect to the horizontal distribution of the soil layers and their composition. We have applied the same procedure as in Kertess case in order to determine the transmissivity. We have discovered a very small variation thus we have used only the average value for our computations.

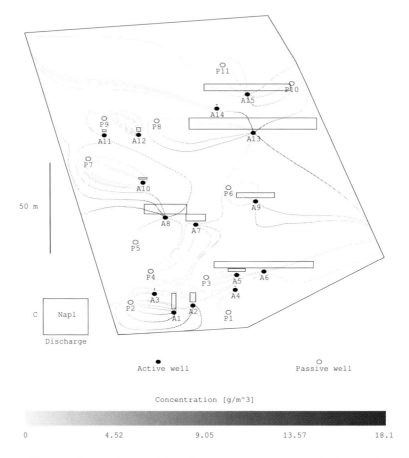

Figure 8.10: Kertess: Optimal tetrachlorethen extraction rates with the total discharge $100 \frac{g}{s}$.

On the other hand the SVE design is different. The extraction wells operate in different groups (the total number of group is 6), i.e., each group is equipped by a separate engine. That means, all wells in one group will have the same differential pressure. The discharge for the first three groups is $300 \frac{m^3}{h}$ (each) and for the others $200 \frac{m^3}{h}$ (each). The total discharge for all extraction wells is $500 \frac{m^3}{h}$, i.e., not all groups can operate simultaneously (maximum 4 group can be in use at the same time).

Data:

- The transmissivity is constant $T = 4 \cdot 10^{-11} \frac{m^2 s^3}{kg}$.

- The volatilization coefficient is constant $V = 4 \cdot 10^{-7} \frac{kg}{m^2 s}$.

- The atmospheric pressure is prescribed as Dirichlet condition on the lateral boundary.

8.2. Kirchweyhe   123

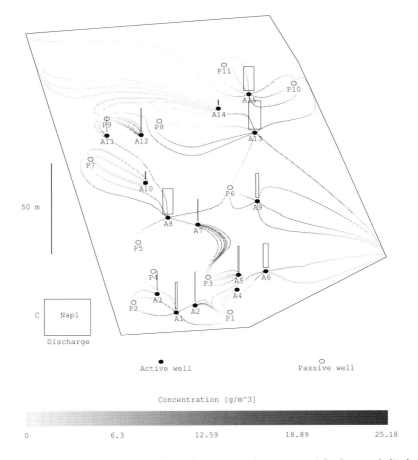

Figure 8.11: Kertess: Optimal tetrachlorethen extraction rates with the total discharge $10\frac{g}{s}$.

- There are 19 active wells divided into 6 groups, shown in Figure 8.14. One cannot prescribe discharges for each well separately, but only pressures or discharges for all wells of a group simultaneously, see the Remark 3.2 at the end of Section 3.1.1.

- The leakage term is $L = 0.01 \ \frac{1}{m^2}$.

8.2.1 Optimizing the Extraction Rates

The situation is different from the previous case and the optimization will have the following new aspects:

- The control variable is the discharges of the groups and not the discharges of the wells.

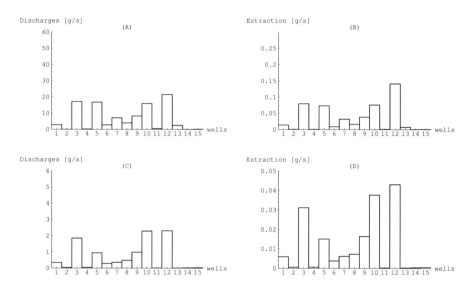

Figure 8.12: Kertess: Optimization of *tetrachlorethen* extraction with positions at the high concentration: (A) Optimal discharges with sum of 100 $\frac{g}{s}$, (B) *Tetrachlorethen* extraction with discharges (A), (C) Optimal discharges with sum of 10 $\frac{g}{s}$ and (D) *Tetrachlorethen* extraction with discharges (C).

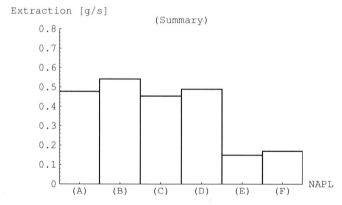

Figure 8.13: Kertess: Extraction rates for *tetrachlorethen* with:
(A) Original discharges and positions, (B) Optimal discharges, original positions and total rate 240 $\frac{g}{s}$, (C) Optimal discharges, original positions and total rate 100 $\frac{g}{s}$, (D) Optimal discharges, better positions and total rate 100 $\frac{g}{s}$, (E) Optimal discharges, original positions and total rate 10 $\frac{g}{s}$, (D) Optimal discharges, better positions and total rate 10 $\frac{g}{s}$.

8.2. Kirchweyhe

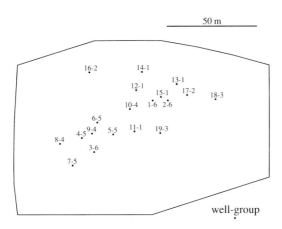

Figure 8.14: Kirchweyhe: Domain and location of active wells.

- We have the following constraints:
  - Only four of the six groups can be used at the same time.
  - For the first three groups, as shown in Figure 8.14, the total extraction of air for each group must not exceed 300 $\frac{m^3}{h}$, while for the second group the limit is 200 $\frac{m^3}{h}$, which corresponds to 107.9 $\frac{g}{s}$ and 71.9 $\frac{g}{s}$ respectively.
  - The total extraction of air of all the groups must not exceed 500 $\frac{m^3}{h}$ which corresponds to 179.8 $\frac{g}{s}$.

In this case we consider the extraction of Trichlorethen (TCE) for its high extraction rate comparing to the other component. The concentration of TCE in the soil before the cleaning procedure is available and we use it to estimate the volatilization coefficient. After a simple calibration of the TCE extraction rate the volatilization coefficient is given in Figure 8.15. This explains the choice of the location of the pumps shown in Figure 8.14. The optimization we have tried follows the following steps.

Elimination of two groups. We first want to determine which four groups of the six we have to use. For all combinations of four groups of the six, we drop the two others and we optimize the discharges. We generate discharges for the groups randomly and we compute the total extraction rates. This first step of optimization finished with the conclusion that the groups number 1 and 2 have to be stopped. This is not surprising when we know that these two groups are not located near the high NAPL sources.

Optimization. We take the best combination with the best realization and we apply Powell's algorithm to improve it. The result is given in Figure 8.16.

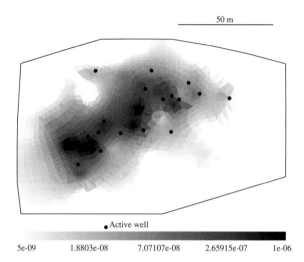

Figure 8.15: Kirchweyhe: Volatilization coefficient for Trichlorethen.

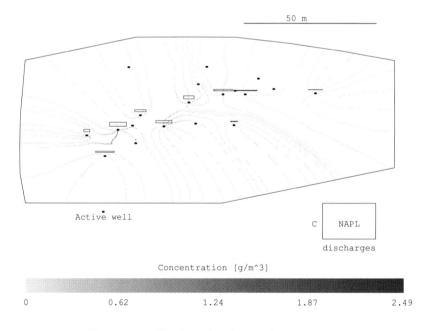

Figure 8.16: Kirchweyhe: Optimal scenario.

Chapter 9

Stochastic Optimization

For simplicity we consider here only Kertess and one single pollutant namely the Tetrachlorethen. The total air pumping rate is also fixed and given the value 100 $\frac{g}{s}$. We suppose that we have two stochastic parameters namely transmissivity and volatilization coefficient. We suppose that the stochastic properties of these variables are known and many realizations of them can be done (see Section 6.3). The Monte Carlo formulation gives a reliable design solution. Let T_i and V_i, $i = 1, \ldots, M$ be M transmissivities and volatilization coefficients generated by Monte Carlo realization. We consider here two aspects of the effect of the spatial variability of these two parameters: the Monte Carlo Optimization and the sensitivity of the average optimum to the parameters.

9.1 Monte Carlo Optimization

For each realization T_i and V_i, $i = 1, \ldots, M$ be M we optimize the pollutant extraction. The number of realizations M has to be defined in function of their results. If the mean and the variance of the solutions tend to stable values then one would stop and consider that the number of realizations is rather sufficient to characterize the stochastic solution. If the mean or the variance continue to vary significantly one should continue and optimize with more realizations.

Figures 9.1, 9.2, 9.3 represent the distribution respectively, mean and standard deviation of the discharges function of the number of realizations. The same three quantities for the extraction rates are shown in Figures 9.4, 9.5, 9.6. Figure 9.7 represents the mean and the standard deviation for the discharges and extraction rates using all the realizations. The discharges are at least correlated by their sum which is constant. The correlation between the discharges and the extraction rates is also visible, cf. Figure 9.7. We observe then that the discharges to the 15 wells for this realization follow some constant behavior. The wells number 1, 2, 3, 4, 10, 11, 12 and 14 should be given rather very low discharges. The wells number 13 and 15 in general should work with high discharges. The rest namely the wells number 5, 6, 7, 8 and 9 have middle discharges. Figure 9.8 represents the mean and the standard deviation optimal total extraction rates of the Monte Carlo realizations.

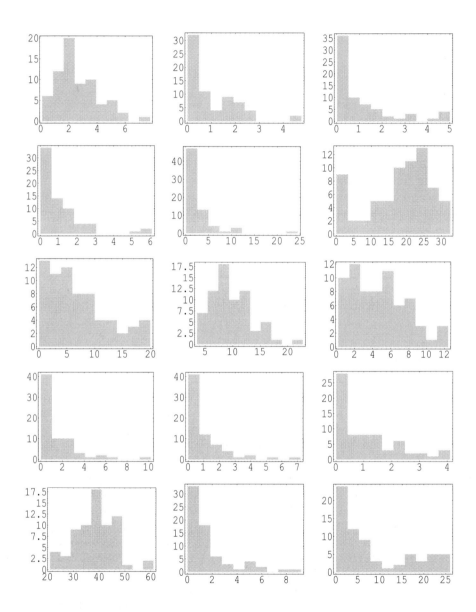

Figure 9.1: Kertess: Distribution function of the discharges $[\frac{g}{s}]$ for 15 wells using 34 realizations.

9.1. Monte Carlo Optimization

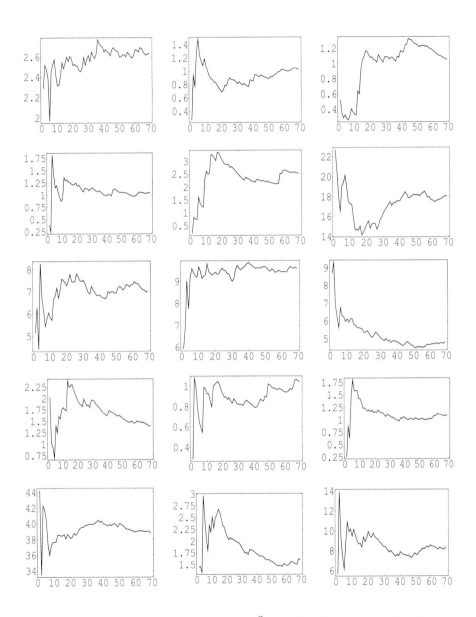

Figure 9.2: Kertess: Mean of the discharges $[\frac{q}{s}]$ for 15 wells function of realizations.

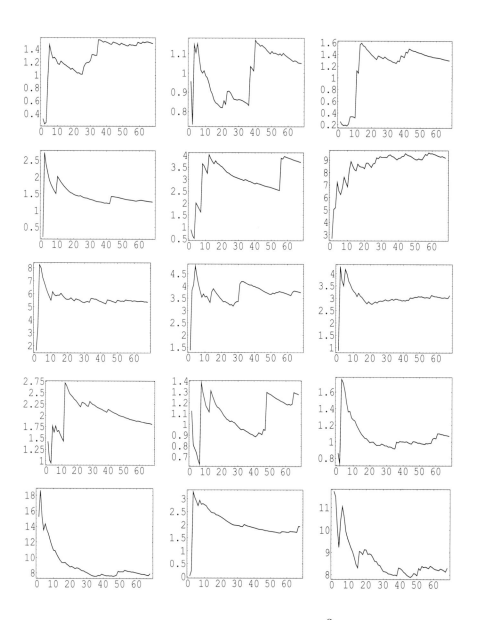

Figure 9.3: Kertess: Standard deviation of the discharges $[\frac{q}{s}]$ for 15 wells function of realizations.

9.1. Monte Carlo Optimization

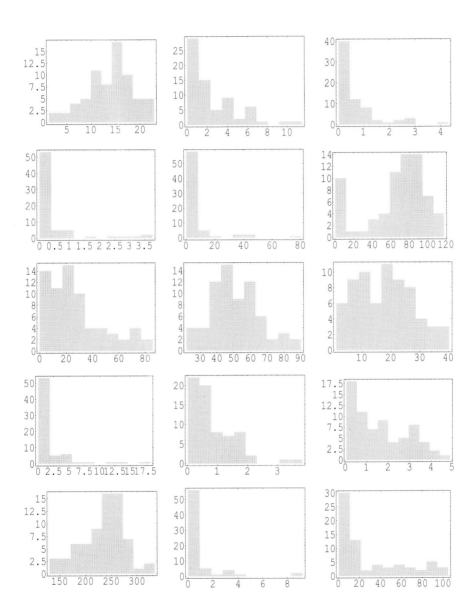

Figure 9.4: Kertess: Distribution function of the napl extraction $[\frac{mg}{s}]$ for 15 wells using 34 realizations.

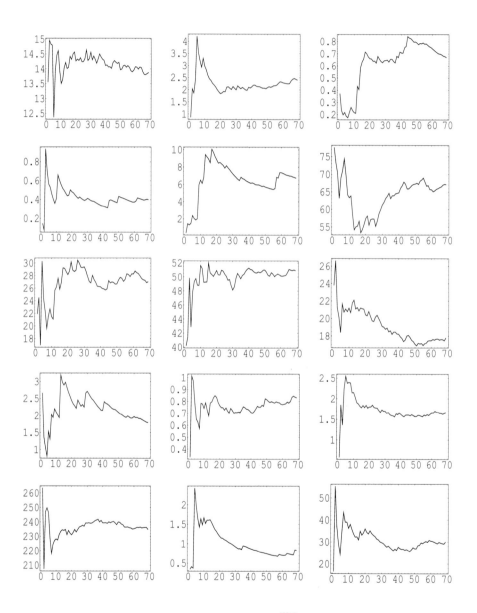

Figure 9.5: Kertess: Mean of the NAPL extraction $[\frac{mg}{s}]$ for 15 wells function of realizations.

9.1. Monte Carlo Optimization

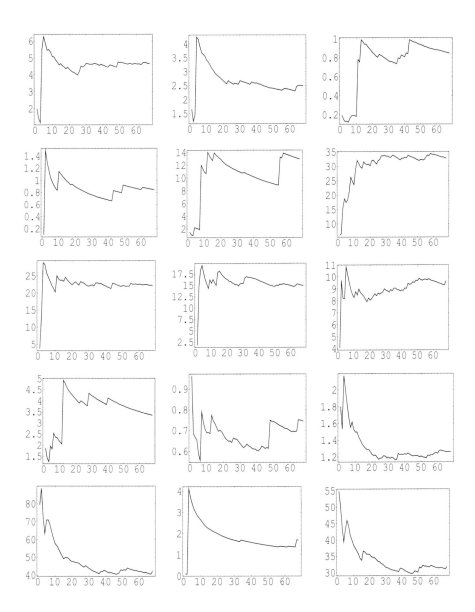

Figure 9.6: Kertess: Standard deviation of the NAPL extraction $[\frac{mg}{s}]$ for 15 wells function of realizations.

9.2 Sensitivity Analysis

Now we take the mean optimum given by Figure 9.7 and compute the pollutant extraction when varying the parameters T_i and V_i in a Monte Carlo method to study the sensitivity of this "optimum" to the spatial variability. It is not true that this mean is always a reliable solution. Figure 9.9 represents the mean and the standard deviation total extraction rates of the Monte Carlo realizations applied to the mean optimal solution given by Figure 9.7. The comparison between Figures 9.8 and 9.9 shows that the mean of optimal solutions is a reliable solution in this case. Indeed, the natural solution for the stochastic problem denoted u^* satisfies:

$$E[J(R, u^*)] \geq E[J(R, u)], \forall u \in U \tag{9.1}$$

where $E[.]$ denotes the expectation with respect to the random variable R and U is the set of acceptable solutions. The Monte Carlo solutions denoted $u_i, i = 1, \ldots, M$ of the last section satisfy

$$J(R_i, u_i) \geq J(R_i, u), \forall u \in U. \tag{9.2}$$

We suppose that M is large enough such that we approximate the mean value of M realization with the expectation value of the corresponding random variable. Therefore, let us denote $\tilde{u} = E[u_i]$. Using equations (9.1) and (9.2), one can easily find out that

$$E[J(R, \tilde{u})] \leq E[J(R, u^*)] \leq E[J(R_i, u_i)]. \tag{9.3}$$

The difference between the first and the last term of the inequality (9.3) is so small as shown in Figures 9.8 and 9.9 that one does not need to look for the solution u^*.

9.2. Sensitivity Analysis

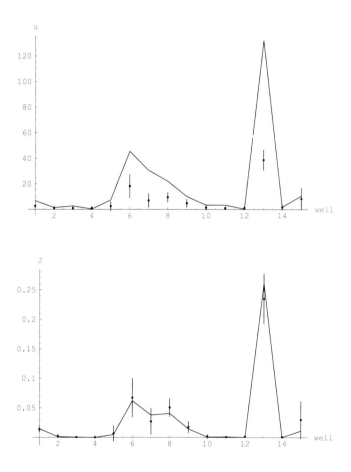

Figure 9.7: Kertess: Arithmetic means of optimal discharges $[\frac{g}{s}]$ (Extraction rates $[\frac{g}{s}]$) for stochastic optimization are denoted by points, standard deviation by segments and the solution with kriging parameters by continuous lines.

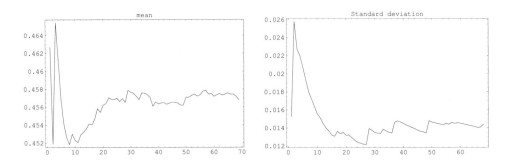

Figure 9.8: Kertess: Mean $[\frac{mg}{s}]$ and standard deviation $[\frac{mg}{s}]$ of the 70 optimal extraction rates solutions of 70 stochastic realizations.

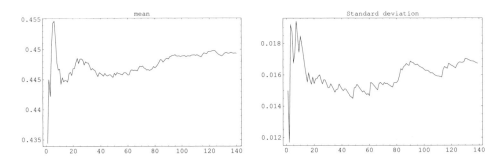

Figure 9.9: Kertess: Mean $[\frac{mg}{s}]$ and standard deviation $[\frac{mg}{s}]$ of 140 extraction rates corresponding to stochastic realizations when using the mean of the optimal discharges.

Chapter 10

Discussions and Conclusions

In this chapter we discuss some aspects and details which have not been mentioned in previous chapters.

Full 3-D Transport Modeling
The modeling of the transport in porous media in 3-D is similar to 2-D. 2-D simplification can be used in some special cases, only. The heterogeneities in the soil matrix, spatial variability of the physical quantities (like pollutant concentration) lead to 3-D considerations. The main problem of computations is the huge number of unknowns (some millions in 3-D case). This is the reason, why some special techniques (like domain decomposition and parallel computing) must be used.

Dual Porosity Models
Dual- or double porosity models assume that the porous medium consists of two different but connected parts (macro-pores or cracks and soil matrix). Fissures and cracks influence the air flow, and on the other hand the soil structure effect is the diffusion from matrix to cracks. Mass transfer in dual-porosity models is frequently approximated by first-order differential equation (cf. Gerke and van Genuchten [GvG93]). Hence dual-porosity models usually involve two flow equations (for macro-pores and soil) coupled by means of a sink/source term describing the transfer between both systems (cf. Hornung [Hor96]).

Field Tests and Tracer Experiments
After the theoretical modeling, one has to test the model for the situation one actually wishes to model. This means, one puts the relevant data into the model and confirms the results with reality. The discrepancies related to uncertainties in the modeling or in the data have to be resolved by model calibration. This can be done using the field measurements of some physical quantities. Sometimes the tracer tests are used in order to get better information about the air flow field in the soil matrix. These need special equipment for measurements.

10.1 Time Dependent Problem

Up to now, we have focused our attention to the steady state problem. Soil remediation is time dependent, i.e., some physical quantities like concentrations, discharges, and others can change in time. This makes the whole modeling and all computations more complicated. We have to distinguish between different time scales.

Time-Scales

Short Time Scale How long does it take for the pressure field to build up? This first period is very short. Hiller [Hil] says that one needs 15–30 minutes to arrive at the steady state regime. Thus, only some minutes are needed to develop a stationary flow field caused by constant differential pressures at the extraction wells.

Intermediate Time Scale How long does it take for the contaminant concentration to become stationary? At the beginning of soil venting the contaminant concentrations and mass removals are high during the first two weeks (cf. Hiller [Hil]). The reason is that NAPLs existing in the equilibrium conditions are removed first. Then the concentrations are going down because of slow phase change of VOCs from the liquid to the air phase.

Long Time Scale How long does it take for the contaminant to be extracted? It can take a period of several months up to some years. This depends on many factors as NAPL concentration, soil properties and optimality of the SVE design.

From the computational point of view, solving a time dependent problem is much more time-consuming than it was by steady state problems. Further, the computational effort increases rapidly by optimization of time dependent problems, which makes the computations tedious.

10.2 Optimization of Well Positions in Large Scale Applications

The optimal control aspect of soil venting consists of optimizing the remediation process with respect to its efficiency. There are several conflicting goals: (i) to maximize the extraction rate of NAPLs, (ii) to minimize the duration of the soil venting process, (iii) to minimize the costs, (iv) respectation of all technical restrictions. It is not obvious how to formulate an appropriate mathematical objective/cost functional taking into account all these items.

One of the most important things is the optimal positioning and the number of the active wells. This depends on many facts: (i) the geometry of the remediation site, (ii) the boundary conditions for the air flow problem, (iii) the geological properties of the soil matrix, (iv) NAPL concentration, (v) available technical restrictions. In some special cases it is possible to solve this problem, e.g., for homogeneous porous medium with one small symmetric contaminated area to put the single active well into the center of contamination. When the effective radius of a single well is much smaller than the diameter of the contaminated domain, one must use more active wells. The praxis is much more complicated (the remediation site is neither homogeneous nor symmetric). There exist some pragmatic algorithms, cf. Kuo, Michel and Gray [KMW92] (for groundwater remediation but for the same problem) and references therein. The existence of a mathematically correct approach which solves the following two questions

- how many active wells are needed
- where to place them

remains an open question.

We suggest the following iterative algorithm how to obtain a relatively reasonable answer to this problem

1. create a relatively dense regular mesh of active wells
2. compute the air flow field and the extraction rates
3. close the "inefficient" probes
4. repeat the points 2 and 3 until the number of active wells becomes reasonable.

10.3 Further Aspects of Stochastic Optimization

We suppose that we have four stochastic parameters namely, transmissivity, volatilization, leakage term and the boundary condition. We assume that the stochastic properties of these variables are known and many realizations of them can be done (see Chapter 9). The Monte Carlo formulation gives a reliable design solution. Let T_i, $i = 1, \ldots, M$ be M transmissivities generated by Monte Carlo realization using (6.34).

Minimizing the Expectation

We propose the following stochastic optimal control problem

$$\inf \{\mathrm{E}\, J(u) : u \in U\}$$

where E denotes the expectations and J is the cost functional defined by equation (3.8) subject to the restrictions given there.

10.3.1 Solution Strategies

There are three possible approaches:

1. Estimate $\inf \mathrm{E}\, J(u)$:

    - Choose as initial value, the control $u \in U$ which minimizes J for the "effective" parameters.

- Determine E $J(u)$ by Monte Carlo realizations.
- Minimize E $J(u)$ over all $u \in U$ using Powell's algorithm.

This procedure is extremely expensive.

2. Estimate E inf $J(u)$, i.e. determine a lower bound of the original objective function:

    - Generate the parameters.
    - Solve the optimal control problem for these parameters.
    - Average over "all" the realizations by repeating this in a Monte Carlo sense.

    This procedure is much less time consuming compared to the first one; the flux and the pressure are affine functions of the control variable for a given set of parameters and the second approach, in opposite to the first one, permits to use this advantage.

3. Use "effective properties":

    - Determine the "effective" (i.e., deterministic) parameters.
    - Solve the optimal control problem for this effective parameters.

    This procedure is cheap, but it does not solve the original problem.

Minimizing the Risk

Another formulation is to consider a problem in which the risk is minimized. Let $j_{10}(u) \in \mathbb{R}$ be chosen such that
$$\pi\{J(u) \geq j_{10}(u)\} = 0.1.$$
Then we say that a control u minimizes the risk at the 10% level, if it solves the problem
$$\inf \{j_{10}(u) : u \in U\} \qquad (10.1)$$

Remark 10.1 If u^* solves (10.1) then for any $u \in U$ one has
$$\pi\{J(u) \geq j_{10}(u^*)\} \geq 0.1 \qquad \square$$

10.3.2 Solution Strategies

1. Estimate inf j_{10}:

    - Choose as initial value, the control $u \in U$ which minimizes J for the "effective" parameters.
    - Determine $j_{10}(u)$ by Monte Carlo realizations.
    - Minimize j_{10} over all $u \in U$ using Powell's algorithm.

    This procedure is extremely expensive.

Bibliography

[AFM94] J. E. Armstrong, U. O. Frind, and R. D. McClellan. Nonequilibrium mass transfer between vapor, aqueous, and solid phases in unsaturated spils during vapor extraction. *Water Resour. Res.*, 30(2):355–368, 1994.

[AP85a] L.M. Abriola and G.F. Pinder. A multiphase approach to the modeling of porous media contamination by organic compounds. 1. Equation development. *Water Resour. Res.*, 21(1):11–18, 1985.

[AP85b] L.M. Abriola and G.F. Pinder. A multiphase approach to the modeling of porous media contamination by organic compounds 2. Numerical simulation. *Water Resour. Res.*, 21(1):19–26, 1985.

[APP93] A.E. Adenekan, T.W. Patzek, and K. Pruess. Modeling of multiphase transport of multicomponent organic contaminants and heat in the subsurface: Numerical model formulation. *Wat. Res. Res.*, (29):3727–3740, 1993.

[AS64] M. Abramowitz and I. A. Stegun. *Handbook of Mathematical Functions*. Dover, 1964.

[BB90] A.L. Baehr and C.J. Bruell. Application of the Stefan-Maxwell equations to determine limitations of Fick's law when modeling organic vapor transport in sand columns. *Water Resour. Res.*, 26(6):1155–1163, 1990.

[BBR79] A.E. Berger, H. Brezis, and J.C.W. Rogers. A numerical method for solving the problem $u_t - \Delta f(u) = 0$. *R.A.I.R.O. Anal. Numer.*, 13:297–312, 1979.

[BC87] A.L. Baehr and M.Y. Corapcioglu. A compositional multiphase model for groundwater contamination by petroleum products 1. Theoretical consideration. *Water Resour. Res.*, 23(1):191–200, 1987.

[BF91] F. Brezzi and M. Fortin. *Mixed and Hybrid Finite Element Methods*. Springer-Verlag, New York, 1991.

[BH91] A.L. Baehr and M.F. Hult. Evaluation of unsaturated zone air permeability through pneumatic tests. *Water Resour. Res.*, 27(10):2605–2617, 1991.

[BHH86] F. Bruckner, H.M. Harress, and D. Hiller. Die Absaugung der Bodenluft – ein Verfahren zur Sanierung von Bodenkontaminationen mit leichtflüchtigen chlorierten Kohlenwasserstoffen. *Brunnenbau, Bau von Wasserwerken, Rohrleitungsbau*, 37(5):3–8, 1986.

[BHM89a]   A. L. Baehr, G. E. Hoag, and M. C. Marley. Removing volatile contaminants from the unsaturated zone by inducing advective air-phase transport. *Journal of Contaminant Hydrology*, 4:1–26, 1989.

[BHM89b]   A.L. Baehr, G.E. Hoag, and M.C. Marley. Removing volatile contaminants from the unsaturated zone by inducing advective air-phase transport. *Journal of Contaminant Hydrology*, (4):1–26, 1989.

[BK85]   F. Bruckner and H. Kugele. Sanierung von mit leichtflüchtigen chlorierten Kohlenwasserstoffen (CKW) kontaminierten Böden mittels Absaugung der Bodenluft. *Korrespondenz Abwasser*, 32(10):863–866, 1985.

[BL73]   K.-F. Busch and L. Luckner. *Geohydraulik*. VEB Deutscher Verlag für Grundstoffindustrie, Leipzig, 1973. 2. Auflage.

[BN92]   J. Bear and J.J. Nitao. Conceptual and Mathematical Modeling of Flow and Contaminant Transport in the Unsaturated Zone Under Nonisothermal Conditions. Technical report, Centro Stefano Franscini-ETH-Zürich, Ascona, 1992. Scientific Colloqium "Porous or Fractured Unsaturated Media: Transport and Behaviour".

[BO91]   I. Braud and Ch. Obled. On the use of empirical orthogonal function (eof) analysis in the simulation of random fields. *Stochastics Hydrology and Hydraulics*, 5:125–134, 1991.

[BPB93]   R. Braun, C. Pennerstorfer, and E. Bauer. Biologische Sanierung kohlenwasserstoffverunreinigter Böden. *Entsorgungs Praxis*, 1(2):25–29, 1993.

[Bre70]   H. Brezis. On some degenerate non-linear parabolic equations. In F.E. Browder, editor, *Nonlinear functional analysis,*, volume 1, pages 28–38. A.M.S. XVIII, 1970.

[Bre73]   R. P. Brent. *Algorithms for Minimization without Derivatives*. Prentice-Hall, New York, 1973.

[Bru91]   M.L. Brusseau. Transport of organic chemicals by gas advection in structured or heterogeneous porous media: development of a model and application to column experiments. *Water Resour. Res.*, 27(12):3189–3199, 1991.

[CAA+92]   V.M. Chen, L.M. Abriola, P.J. Alvarez, P.J. Anid, and T.M. Vogel. Modeling transport and biodegradation of benzene and toluene in sandy aquifer material: comparison with experimental measurements. *Water Resour. Res.*, 28(7):1833–1847, 1992.

[CB87]   Y.M. Corapcioglu and A.L. Baehr. A compositional multiphase model for groundwater contamination by petroleum products. 1. Theoretical considerations. *Water Resour. Res.*, 23(1):191–200, 1987.

Bibliography

[CB92]   M.A. Celia and P. Binning. A mass conservative numerical solution for two-phase flow in porous media with application to unsaturated flow. *Water Resour. Res.*, 28(10):2819–2828, 1992.

[CBUS92]   S. Coffa, J. Boode, L.G.C.M. Urlings, and F. Spuy. Physikalische Beschreibung des Massentransportes bei Bodenluftextraktion. *WLB Wasser, Luft und Boden*, (4):73–76, 1992.

[CKS90]   J. Croisé, W. Kinzelbach, and J. Schmolke. Luftströmungen in der ungesättigten Bodenzone bei In-Situ-Sanierungsmaßnahmen, Wasser und Boden. *Wasser und Boden*, 42(3):151–170, 1990.

[CL71]   M.G. Crandall and T.M. Liggett. Generation of semigroups of nonlinear transformations on general banach spaces. *Amer. J. Math.*, 93:265–298, 1971.

[CL91]   P.G. Ciarlet and J.-L. Lions. *Finite Element Methods*, volume II of *Handbook of Numerical Analysis*. Horth-Holland, Amsterdam, 1991.

[CSL91]   T.B. Culver, C.A. Shoemaker, and L.W. Lion. Impact of vapor sorption on the subsurface transport of volatile organic compounds: a numerical model and analysis. *Water Resour. Res.*, 27(9):2259–2270, 1991.

[CW80]   R.E. Cunningham and R.J. Williams. Diffusion in gases and porous media. *Plenum, New York*, page 275 pp., 1980.

[Del78]   J. P. Delhomme. Kriging in the hydrosciences. *Adv. Wat. Res.*, 1:251–266, 1978.

[DIN87]   DIN 4022, Benennen und Beschreiben von Boden und Fels, 1987. Deutsches Institut für Normung e. V., Berlin.

[DL90]   R. Dautray and J.-L. Lions. *Physical Origins and Classical Methods*, volume 1 of *Mathematical Analysis and Numerical Methods for Science and Technology*. Springer, Berlin, Heidelberg, 1990.

[DL93]   R. Dautray and J.-L. Lions. *Evolution Problems II*, volume 6 of *Mathematical Analysis and Numerical Methods for Science and Technology*. Springer, Berlin, Heidelberg, 1993.

[DMW92]   S.G. Donaldson, G.C. Miller, and Miller W.W. Remediation of gasoline-contaminated soil by passive volatilization. *J. Environ. Qual.*, 21(1):94–102, 1992.

[DR91]   A.H. Demond and P.V. Roberts. Effect of interfacial forces on two-phase capillary pressure-saturation relationships. *Water Resour. Res.*, 27(3):423–437, 1991.

[ea91]   Tietje et al. Ptf-code. Report, Technical University Braunschweig, 1991.

[EGS90] G. Einsele, P. Grathwohl, and M. Sanns. Verteilung und Ausbreitung verschiedener leichtflüchtiger chlorierter Kohlenwasserstoffe (LCKW) im System Feststoff-Bodenwasser-Bodenluft (und Übergang ins Grundwasser). In P. Bock, H. Hötzl, and M. Nahold, editors, *Untergrundsanierung mittels Bodenluftabsaugung und In-Situ-Strippen*, pages 33–50. 1990. Schriftenreihe Angewandte Geologie Karlsruhe 9 I-VII.

[Fis95] U. Fischer. Experimental and numerical investigation of soil vapor extraction. Ph.D. dissertation, ETH. Zürich No. 11277, 1995.

[FJPW89] R.W. Falta, I. Javandel, K. Pruess, and P.A. Witherspoon. Density-driven flow of gas in the unsaturated zone due to the evaporation of volatile organic chemicals in columns of unsaturated soil. *Water Resour. Res.*, 25(10):2159–2169, 1989.

[FPJW92] R.W. Falta, K. Pruess, I. Javandel, and P.A. Witherspoon. Numerical modeling of steam injection for the removal of nonaqueous phase liquids from the subsurface 1. Numerical formulation. *Water Resour. Res.*, 28(2):433–449, 1992.

[FS91] P.A. Forsyth and B.Y. Shao. Numerical simulation of gas venting for NAPL site remediation. *Adv. Wat. Res.*, 14(6):354–367, 1991.

[FY93] P. Fine and B. Yaron. Outdoor experiments on enhanced volatilization by venting of kerosene component from soil. *J. Contam. Hydrol.*, 12(4):355–370, 1993.

[GHC90] J.S. Gierke, N.J. Hutzler, and J.C. Crittenden. Modeling the movement of volatile organic chemicals in columns of unsaturated soil. *Water Resour. Res.*, 26(7):1529–1547, 1990.

[GHM92] J.S. Gierke, N.J. Hutzler, and D.B. McKenzie. Vapor transport in unsaturated soil columns: implications for vapor extraction. *Water Resour. Res.*, 28(2):323–33, 1992.

[Gim94] T.F. Gimmi. Transport of reactive gases in unsaturated, structured media – modeling and experimental aspects. Ph.D. dissertation, ETH. Zürich No. 10904, 1994.

[GKLS91] G.W. Gee, C.T. Kincaid, R.J. Lennard, and C.S. Simmons. Recent studies of flow and transport in the vadose zone. *Reviews of Geophysics, Supplement*, pages 227–239, 1991.

[Gol89] D. E. Goldberg. *Genetic Algorithms in Search, Optimization, and Machine Learning*. Addison-Wesley Publishing Company, INC, New York, 1989.

[Gor90] S. M. Gorelick. Large scale nonlinear deterministic and stochastic optimization: Formulations involving simulation of subsurface contamination. *Mathematical Programming*, 48:19–39, 1990.

[GR80]   I.S. Gradshteyn and I.M. Ryzhik. *Table of Integrals, Series, and Products*. Academic Press, 1980.

[Gra90]   P. Grathwohl. Bestimmung der Gesamtbelastung des Bodens mit leichtflüchtigen Schadstoffen aus der Schadstoffkonzentration in der Bodenluft. *Wasser und Boden*, 42(4):244–250, 1990.

[GS80]   I. I. Gihmann and A. V. Skorohod. *The Theory pf Stochastic Processes Vol I*. Springer, 1980.

[GT83]   D. Gilbarg and N. S. Trudinger. *Elliptic Partial Differential Equations of Second Order*. Springer, Berlin, Heidelberg, 1983.

[GvG93]   H.H. Gerke and M.Th. van Genuchten. A dual-porosity model for simulating the preferential movement of water and solutes in structured porous media. *Water Resour. Res.*, 29(2):305–319, 1993.

[HGK$^+$97] U. Hornung, H. Gerke, Y. Kelanemer, S. Schumacher, M. Slodička, and D. Stegemann. Optimierung von Anlagen zur Bodenluftabsaugung. In K.-H. Hoffmann, W. Jäger, T. Lohmann, and H. Schunck, editors, *Mathematik – Schlüsseltechnologie für die Zukunft*, pages 205–218, Berlin, Heidelberg, New York, 1997. Springer.

[Hil]   D.H. Hiller. Performance charasteristics of vapor extraction systems operated in Europe. In *Proceedings of the symposium on soil venting, Houston, Texas, 1991*, pages 193–202. Enviromental Protection Agency. EPA/600/R-92/174.

[HKS96]   U. Hornung, Y. Kelanemer, and M. Slodička. Soil venting. In K. Malanowski, Z. Nahorski, and M. Peszyńska, editors, *Modelling and Optimization of Distributed Parameter Systems with Applications to Engineering*. Chapman & Hall, 1996.

[HMS90]   H.M. Harress, K.-H. Munz, and T. Schöndorf. Bodenluftabsaugung. Weber H. H. (Ed.) "Altlasten. Erkennen. Bewerten. Sanieren":328–347, 1990. Springer Verlag, Berlin.

[Hor92]   U. Hornung. Modellierung von Stofftransport in aggregierten und geklüfteten Böden. *Wasserwirtschaft*, 92(1), 1992.

[Hor96]   U. Hornung. *Homogenization and Porous Media*, volume 6 of *Interdisciplinary Applied Mathematics*. Springer, 1996.

[HP86]   R. Haverkamp and J.-Y. Parlange. Predicting the water-retention curve from particle-size distribution: 1. sandy soils without organic matter. *Soil Science*, 142(6):325–339, 1986.

[HS92]   R. Homringhausen and M. Schwarz. Bohrarbeiten für die Sanierung einer Schlammentwässerungsanlage. *bbr – Brunnenbau, Bau von Wasserwerken, Rohrleitungsbau*, 43(11):480–48, 1992.

[JH78]   A. G. Journel and Ch. J. Huijbregts. *Mining Geostatistics*. Academic Press, 1978.

[JK91]   W. Jäger and J. Kačur. Solution of porous medium type systems by linear approximation schemes. *Numer. Math.*, 60:407–427, 1991.

[KF91a] B.H. Kueper and E.O. Frind. Two-phase flow in heterogeneous porous media, 1.Model development. *Water Resour. Res.*, 27(6):1049–1057, 1991.

[KF91b] B.H. Kueper and E.O. Frind. Two-phase flow in heterogeneous porous media, 2. Model application. *Water Resour. Res.*, 27(6):1059–1070, 1991.

[KH90]   E. F. Kaasschieter and A. J. M. Huijben. Mixed hybrid finite element and streamline computation for the potential flow problem. Technical Report Report PN 90-92-A, TU Delft, 1990.

[KJF77] A. Kufner, O. John, and S. Fučík. *Function spaces*. Academia, Prague, 1977.

[KMW92] Ch.-H. Kuo, A.N. Michel, and Gray W.G. Design of optimal pump-and-treat strategies for contaminated groundwater remediation using the simulated annealing algorithm. *Adv. Wat. Res.*, 15:95–105, 1992.

[KT75]   S. Karlin and H. M. Taylor. *A First Course in Stochastic Processes*. Academic Press, 1975.

[Küh90] W. Kühn. Beurteilung und Behandlung von Verschmutzungs-Situationen mit chlorierten Kohlenwasserstoffen. In P. Bock, H. Hötzl, and M. Nahold, editors, *Untergrundsanierung mittels Bodenluftabsaugung und In-Situ-Strippen*, pages 27–32. 1990. Schriftenreihe Angewandte Geologie Karlsruhe 9 I-VII.

[Lan92] A. Lange. Permeabilitätsmessungen von Bodenproben vom Kertess-Gelände/Hannover. Institut für Wasserwirtschaft, Hydrologie und Landwirtschaftlichen Wasserbau, Universität Hannover, Report for GEO-data, 1992.

[LD92]   S. Lingineni and V.K. Dhir. Modeling of soil venting processes to remediate unsaturated soils. *J. Environ. Eng.*, 118(1):135–151, 1992.

[Lio71] J.-L. Lions. *Optimal Control of Systems Governed by Partial Differential Equations*. Springer, Berlin, Heidelberg, 1971.

[LP87a] R.J. Lenhard and J.C. Parker. A model for hysteretic constitutive relations governing multiphase flow 2. Permeability-saturation relations. *Water Resour. Res.*, 23(12):2197–2206, 1987.

[LP87b] R.J. Lenhard and J.C. Parker. A model for hysteretic constitutive relations governing multiphase flow 1. Saturation-pressure relations. *Water Resour. Res.*, 23(12):2187–2196, 1987.

Bibliography

[LPK91]   R.J. Lenhard, J.C. Parker, and J.J. Kaluarachchi. Comparing simulated and experimental hysteretic two-phase transient fluid flow phenomena. *Water Resour. Res.*, 27(8):2113–2124, 1991.

[Mar87]   M. J. Maron. *Numerical Analysis, A Practical Approch*. Macmillan Publishing Company, New York, 1987.

[MC90]   J.W. Mercer and R.M. Cohen. A review of immiscible fluids in the subsurface: Properties, models, characterization and remediation. *J. Contam. Hydrol.*, (6):107–163, 1990.

[Mey63]   N.G. Meyers. An L_p-estimate for the gradient of solutions of second order elliptic divergence equations. *Ann. Sc. Norm. Sup. Pisa*, 17:189–206, 1963.

[MF90a]   C.A. Mendoza and E.O. Frind. Advective-dispersive transport of dense organic vapors in the unsaturated zone. 1. Model development. *Water Resour. Res.*, 26(3):379–387, 1990.

[MF90b]   C.A. Mendoza and E.O. Frind. Advective-dispersive transport of dense organic vapors in the unsaturated zone. 2. Sensitivity analysis. *Water Resour. Res.*, 26(3):388–398, 1990.

[MF92]   J. Massmann and D.F. Faffier. Effects of atmospheric pressures on gas transport in the vadose zone. *Water Resour. Res.*, (28):777–791, 1992.

[MM83]   E.A. Mason and A.P. Malinauskas. Gas Transport in Porous Media: The Dusty-Gas Model. *Chem. Eng. Monogr.*, (17):194 pp., 1983. Elsevier, New York.

[MME67]   E.A. Mason, A.P. Malinauskas, and R. D. Evans. Flow and Diffusion of Gases in Porous Media. *J. Chem. Phys.*, (46):3199–3216, 1967.

[MMP92]   R. Mull, J. Mull, and M. Pielke. Wirkung der Sanierung von CKW Schadensfällen auf die Grundwasserqualität und zugeordnete Kosten – am Beispiel der Stadt Hannover. *Wasser und Boden*, 44(8):510–515, 1992.

[MNV87]   E. Magenes, R.H. Nochetto, and C. Verdi. Energy error estimates for a linear scheme to approximate nonlinear parabolic equations. *R.A.I.R.O Modél. Math. Numér.*, 21:655–678, 1987.

[MW82]   A. Mantoglou and J.L. Wilson. The turning bands method for simulation of random fields using line generation by a spectral method. *Water Resources Research*, 18:1379–1394, 1982.

[NG91]   M. Nahold and K. Gottheil. CKW-Schadensfälle – Die Optimierung von Bodenluftabsaugungen. *WLB Wasser, Luft und Boden*, 11(12):83–86, 1991.

[NS76]   J.A. Nohel and D.F. Shea. Frequency domain methods for Volterra equations. *Advances in Mathematics*, 22:278–304, 1976.

[NZV95]   R. Nieuwenhuizen, W. Zijl, and M. Van Veldhuizen. Flow pattern analysis for a well defined by point sinks. *Transport in Porous Media*, 21:209–223, 1995.

[OK91]   D.W. Ostendorf and D.H. Kampbell. Biodegradation of hydrocarbon vapors in the unsaturated zone. *Water Resour. Res.*, 27(4):453–462, 1991.

[PK92]   M.M. Poulsen and B.H. Küper. A field experiment to study the behavior of tetrachloroethylene in unsaturated porous media. *Environ. Sci. Technol.*, 26(5):889–895, 1992.

[PL87]   J.C. Parker and R.J. Lenhard. A model for hysteretic constitutive relations governing multiphase flow 1. saturation-pressure relations. *Water Resour. Res.*, 23(12):2187–2196, 1987.

[PLAWJ91]   S.E. Powers, C.O. Loureiro, L.M. Abriola, and W.J. Weber Jr. Theoretical study of the significance of nonequilibrium dissolution of nonaqueous phase liquids in the subsurface systems. *Water Resour. Res.*, 27(4):463–477, 1991.

[PLK87]   J.C. Parker, R.J. Lenhard, and T. Kuppusamy. A parametric model for constitutive properties governing multiphase flow in porous media. *Water Resource Research*, 23(4):618–624, 1987.

[PS93]   M. Pantazidou and N. Sitar. Emplacement of nonaqueous liquids in the vadose zone. *Water Resour. Res.*, 29(3):705–722, 1993.

[QR92]   K.-P. Quanz and C. Röhr. Ökobilanz der Bodensanierung durch Bodenluftabsaugung. *WLB Wasser, Luft und Boden*, 1(2):72–74, 1992.

[Reh92]   U. Rehnig. Entfernung leichtflüchtiger organischer Verunreinigungen aus Boden und Grundwasser. *bbr – Brunnenbau, Bau von Wasserwerken, Rohrleitungsbau*, 43(4):149–153, 1992.

[Sch84]   F. Schwille. Leichtflüchtige Chlorkohlenwasserstoffe in porösen und klüftigen Medien-Modellversuchen. *Besond. Mitt. Deutsch. Gewässerkundl. Jahrb.*, (46):72+XII, 1984.

[Sch88]   F. Schwille. *Dense chlorinated solvents in porous and fractured media*. Lewis Publishers, Chelsea, Michigan, USA, 1988.

[SFJ92]   KC. Shan, R.W. Falta, and T. Javandel. Analytical solutions for steady state gas flow to a soil vapor extraction well. *Water Resour. Res.*, 28(4):1105–1120, 1992.

[Sim72]   Ch.G. Simader. *On Dirichlet's boundary value problem*, volume 268 of *Lecture Notes in Math*. Springer, Berlin·Heidelberg, 1972.

[SJ72]   M. Shinozuka and C.-M. Jan. Digital simulation of random processes and its applications. *Journal of Sound and Vibration*, 25:111–128, 1972.

Bibliography

[Slo97]   M. Slodička. Mathematical treatment of point sources in a flow through porous media governed by Darcy's law. *Transport in Porous Media*, 28(1):51–67, 1997.

[Slo98]   M. Slodička. Finite elements in modeling of flow in porous media: How to describe wells. *AMUC*, LXVII(1):197–214, 1998.

[SMY96]   Y.-H. Sun, Davert M.W., and W.W.-G. Yeh. Soil vapor extraction system design by combinatorial optimization. *Water Resources Research*, 32(6):1863–1873, 1996.

[SS89]   B.E. Sleep and J.F. Sykes. Modeling the transport of volatile organics in variably saturated media. *Water Resour. Res.*, 25(1):81–92, 1989.

[SS96]   S. Schumacher and M. Slodička. Bacterial growth and bioremediation. In F. Keil, W. Mackens, H. Voß, and J. Werther, editors, *Scientific computing in chemical engineering*, pages 205–211. Springer, 1996.

[SS97]   S. Schumacher and M. Slodička. Effective radius and underpressure of an air extraction well in a heterogeneous porous medium. *Transport in Porous Media*, 29(3):323–340, 1997.

[Sta66]   G. Stampacchia. *Équations elliptiques du second ordre à coefficients discontinus*. Les Presses de l'Université de Montréal, Montréal, 1966.

[Sta76]   O. J. Staffans. An inequality for positive definite Volterra kernels. *Proc. American Math. Society*, 58:205–210, 1976.

[SX93]   B.A. Schrefler and Zhan X. A fully coupled model for water and airflow in deformable porous media. *Water Resour. Res.*, 29(1):155–167, 1993.

[Tho77]   J. M. Thomas. *Sur l'analyse numérique des méthodes d'éléments finis hybrides et mixtes*. PhD thesis, University Pierre et Marie Curie, Paris, 1977.

[TM92]   C. D. Travis and J. MacInnis. Vapor extraction of organics from subsurface soils, Is it effective? *Environ. Sci.Technol.*, 26(10):1885–1887, 1992.

[TP89a]   D.C. Thorstenson and D.W. Pollock. Gas transport in unsaturated porous media: Multicomponent systems and the adequacy of Fick's law. *Water Resour. Res.*, 25(3):477–507, 1989.

[TP89b]   D.C. Thorstenson and D.W. Pollock. Gas transport in unsaturated porous media: The adequacy of Fick's law. *Rev. Geophys.*, 27(1):61–78, 1989.

[van80]   M.Th. van Genuchten. A closed-form for predicting the hydraulic conductivity of unsaturated soils. *Soil Sci. Soc. Am. J.*, 44:892–898, 1980.

[vGLY91]   M. T. van Genuchten, F.J. Leij, and S.R. Yates. The retc-code for quantifying the hydraulic functions of unsaturated soils. Report EPA/600/2-91/065, Washington, DC, 1991.

[Wil95] D.J. Wilson. *Modeling of insitu techniques for treatment of contaminated soils: Soil vapor extraction, sparging, and bioventing.* Technomic Publishing Company, Inc., Lancester-Basel, 1995.

[ZP93] D.G. Zeitoun and G.F. Pinder. An optimal control least squares method for solving coupled flow-transport systems. *Water Resour. Res.*, 29(2):217–227, 1993.

Index

active wells, 16
adaptive quadrature integration
    rule, 105
adjoint problem, 37
air permeability, 67, 86
air transmissivity, 86

Bessel
    differential equation, 28
    function, modified, 36
    functions, 28
Biological decay, 11
Borel measures
    space of, 37

calibration, 88
capillarity, 9
Chernoff formula
    nonlinear, 13
Clausius-Clapeyron equation, 10
cleaning procedure, 115
condition
    discharge, 18
    flux, 18
    pressure, 17
conjugate gradient method, 103
contaminants, 7
control variables, 19
cost functional, 19
covariance matrix, 73
Crandall-Liggett formula, 13
crossover operator, 108

dimensionless form, 20
discharge condition, 18
dispersion, 9
distribution
    grain-size, 60
    particle-size, 60

dual-porosity system, 14

effective
    grain diameter, 66
    particle-size diameter, 62
    radius, 15, 29
    saturation, 67
estimation procedure, 59
Evaporation, 10
extraction rate, 31
extraction wells, 114

fitness function, 107
flow effective porosity, 62
flux condition, 18
Fourier representation, 81
Fourier transform, 80
fundamental solution, 29

genetic algorithm, 106, 107
grain-size distribution, 60
groups of wells, 100

Hazen's approximation, 63
Henry's law, 10
Hölder continuity, 35
hysteresis, 11

ideal gas law, 95
inf-sup condition, 102

Kertess test-site, 111
kinematic viscosity, 63
kinetics model, 9
Kirchweyhe test-site, 121
Klinkenberg effect, 10, 68
Knudsen diffusion, 10
Kozeny-Carman equation, 62
kriging, 73

Lagrange multipliers, 101

leakage function for the boundary, 17
leakage term, 17
Lipschitz continuous boundary, 37
long-term contaminations, 8

macro-pore system, 14
method of superposition, 37
mixed hybrid finite element method, 100
mixed variational formulation, 101
model of Brooks and Corey, 67
models of van Genuchten and Mualem, 67
modified Bessel function, 36
Monte Carlo realization, 127
Monte Carlo simulations, 77
multi-phase transport models, 11
mutation operator, 108

Newton methods, 107
non-equilibrium conditions, 9
nonlinear Chernoff formula, 13
nonlinear semigroups of contractions
    theory of, 13

optimal control problem, 20
optimization of well positions, 138
optimized extraction rate, 116

particle-size distribution, 60
passive wells, 16, 114
permeability
    air, 67, 86
    coefficient, 62
    relative, 67
    specific, 62
point sinks, 24
population, 107
porosity, 62
positive definite, 77
Powell's algorithm, 106, 108
pressure condition, 17
pressure head, 67
prolongation operator, 37

range of influence, 15
Raoult's law, 10
RED-GREEN algorithm, 110

reducing the total air discharge rate, 117
regular triangulation, 102
relative permeability, 67
reproduction, 107
residual water saturation, 63

saturated hydraulic conductivity, 67
second calibration, 91
semi-variogram, 72, 76
sensitivity analysis, 134
Simpson's rule, 105
soil moisture, 8
solution
    very weak, 38
    weak, 35
space of Borel measures, 37
specific permeability, 62
Stampacchia
    transposition method of, 37
stationary random field, 73
streamline, 30, 104
streamline method, 30
strongly positive definite, 77
subtraction of singularity, 40
SVE design, 122
symmetric configurations of the wells, 53

theory of nonlinear semigroups of contractions, 13
thermodynamical equilibrium, 8
time scales, 138
tortuosity, 62
transposition method of Stampacchia, 37
turning band method, 79

variations of constants, 49
venting technique, 1
very weak solution, 38
volatile organic compounds, 7
volatilization, 16
    coefficient, 94, 113
    rate, 8

weak solution, 35